"A beautifully illustrated digital archeology of what some call 'the first Industrial city' – Manchester. Francesca Froy charts the evolution and integration, the complexity and morphology of a place made by people and the materials they consume, process and rely on. Relying on a wider range of sources than almost any other book of this kind, the result is an academic masterpiece."

Danny Dorling, *Professor, University of Oxford*

"This is a much needed publication. It takes on the mantle of Jane Jacobs, discussing urban complexity in a way that will appeal to people who are interested in actual cities rather than cities in the abstract. It is written with a lightness of touch which is not easily achieved, and which will appeal to a wide readership."

Sam Griffiths, *Bartlett School of Architecture, University College London*

"We are all captivated by complexity, recognising its pervasive presence in our world and striving to understand it. Many dream of mastering it or at least guiding it towards desired outcomes. Yet, complexity remains elusive, especially within urban environments. Francesca Froy presents a fresh and insightful perspective on deciphering the intricate fabric of cities and their local economies. Her configurational view explores how economies unfold and self-organise into specialised branches through the spatial topology of cities. With a captivating style, Froy unveils the hidden threads that shape urban life, offering readers a compelling journey into the heart of urban complexity."

Vinicius Netto, *University of Porto, Portugal*

"This book moves beyond the rhetoric of 'resilient cities' to show how the economy of cities is geared to urban form and street networks at multiple scales. In a masterful exploration of urban complexity and self-organization, Froy shows how cities work to sustain livelihoods, how planning can damage them, and how post-industrial cities can be sustained."

Kim Dovey, *Chair of Architecture and Urban Design and Co-Director of the Informal Urbanism Research Hub, University of Melbourne*

"This book is exceptionally clear and accessible, with a biographic tone that engages the reader. By integrating system theory into a comprehensive urban theoretical framework, it makes complex concepts understandable and relevant for both practitioners and academics."

Patricia Canelas, *Sustainable Urban Development Programme, University of Oxford*

"This important book challenges our conceptions of top-down, deterministic urban spatial planning and economic policy making in cities. With a strong focus on the self-organising complexity and catalytic nature of cities, the author has woven a scholarly, powerful and unique tapestry of interlocking and interdisciplinary concepts. This is strongly grounded in a detailed and insightful case study of networks and urban configurations in Greater Manchester, supported by further discussion of Sheffield, Newcastle and New Haven. Using a variety of innovative data, the ebb and flow, and evolution and destruction in cities over time is brought to life in a highly readable and engaging way. The book offers powerful discussions and visual representations of space and economy (and of complexity and networks), as well as valuable policy and practice recommendations, which academics, students, policymakers and practitioners will find immensely helpful in aiding a clearer understanding of how cities really 'work'."

Timothy J. Dixon, *Emeritus Professor, University of Reading,*
and Visiting Fellow, Kellogg College

"In *Rebuilding Urban Complexity*, Francesca Froy offers a compelling exploration of how cities evolve, decline, and reinvent themselves. Blending insights from complexity science, economics, and urban planning, Froy is able to connect the dots between how cities are built and how their economies work. Using Greater Manchester as a primary case study, alongside other post-industrial cities, she uncovers the intricate networks and interdependencies that shape urban economies and spaces over time. Froy's analysis of historical branching processes and configuration of street networks provides crucial insights for understanding urban resilience and adaptation. Her interdisciplinary approach, combining quantitative network analysis with rich qualitative detail, offers a fresh and exciting framework for conceptualizing cities as complex evolving systems.

This book is essential reading for urban planners, policy makers, and scholars seeking to understand the dynamics of post-industrial cities. Froy's work sheds new light on what makes cities tick and offers valuable lessons for rebuilding vibrant, resilient urban economies and spaces."

Elvira Uyarra, *Professor of Innovation Studies and Director*
of Manchester Institute of Innovation Research

Rebuilding Urban Complexity

This is a book about urban complexity – how it evolves and how it gets destroyed. It explores the structures of interdependency which underpin cities, where the many different "parts" (people, streets, industry sectors) interact to form an evolving "whole".

The book explores the evolution and destruction of complexity in one city – Greater Manchester – but also other post-industrial cities, including Sheffield and Newcastle, Detroit and New Haven. The focus is on the *networked* qualities of public urban space, and how street networks work as multiscale systems. The book also explores economic networks, and the evolving sets of interconnecting economic capabilities which help to shape urban economies. It demonstrates how cities evolve through processes of self-organisation – and concludes by considering how policy makers can best harness such processes as they rebuild urban complexity following insensitive planning interventions in the 1960s and 1970s.

The book will appeal to anybody with an interest in cities, and how they work. It is interdisciplinary in scope, weaving in strands from architecture, economics, history, anthropology and ecology. It is written for academics but also non-academics, including urban planners, architects, local economic development actors and other policy makers.

Francesca Froy is Lecturer on Sustainable Urban Development at the University of Oxford and is a Fellow of Kellogg College. She has honorary positions at the Bartlett Schools of Planning and Architecture where she previously researched and taught urban morphology and local economic development. She is also an associate working on the spatial dimensions of sustainable economies at the consultancy firm Space Syntax. Francesca was a senior policy analyst at the Organisation for Economic Cooperation and Development (OECD) from 2005 to 2015, where she coordinated international reviews of policies to support local economic development. Before that, while based in Brussels, she evaluated urban and regional European policies. Her articles can be found in peer-reviewed academic journals including the *Cambridge Journal of Regions, Economy and Society; Local Economy; the Oxford Review of Economic Policy; European Planning Studies; the Journal of Urban Design;* and *Environment and Planning A: Economy and Space.*

Routledge Research in Planning and Urban Design

Routledge Research in Planning and Urban Design is a series of academic monographs for scholars working in these disciplines and the overlaps between them. Building on Routledge's history of academic rigour and cutting-edge research, the series contributes to the rapidly expanding literature in all areas of planning and urban design.

Inclusion and Exclusion of the Urban Poor in Dhaka
Power, Politics and Planning
Rasheda Rawnak Khan

Remodelling to Prepare for Independence
The Philippine Commonwealth, Decolonisation, Cities and Public Works,
c. 1935–46
Ian Morley

Co-Creative Placekeeping in Los Angeles
Artists and Communities Working Together
Brettany Shannon, David C. Sloane and Anne Bray

Australia and China Perspectives on Urban Regeneration and Rural Revitalization
Raffaele Pernice and Bing Chen

Contested Airport Land
Social-Spatial Transformation and Environmental Injustice in Asia and Africa
Edited by Irit Ittner, Sneha Sharma, Isaac Khambule and Hanna Geschewski

Rebuilding Urban Complexity
A Configurational Approach to Postindustrial Cities
Francesca Froy

For more information about this series, please visit: www.routledge.com/Routledge-Research-in-Planning-and-Urban-Design/book-series/RRPUD

Rebuilding Urban Complexity
A Configurational Approach
to Postindustrial Cities

Francesca Froy

LONDON AND NEW YORK

Designed cover image: photo of warehouse buildings in central Manchester by Surya Prasad on Unsplash.

First published 2025
by Routledge
4 Park Square, Milton Park, Abingdon, Oxon OX14 4RN

and by Routledge
605 Third Avenue, New York, NY 10158

Routledge is an imprint of the Taylor & Francis Group, an informa business

© 2025 Francesca Froy

The right of Francesca Froy to be identified as author of this work has been asserted in accordance with sections 77 and 78 of the Copyright, Designs and Patents Act 1988.

All rights reserved. No part of this book may be reprinted or reproduced or utilised in any form or by any electronic, mechanical, or other means, now known or hereafter invented, including photocopying and recording, or in any information storage or retrieval system, without permission in writing from the publishers.

Trademark notice: Product or corporate names may be trademarks or registered trademarks, and are used only for identification and explanation without intent to infringe.

British Library Cataloguing-in-Publication Data
A catalogue record for this book is available from the British Library

Library of Congress Cataloging-in-Publication Data
Names: Froy, Francesca, author.
Title: Rebuilding urban complexity : a configurational approach to
     postindustrial cities / Francesca Froy.
Description: Abingdon, Oxon ; New York, NY : Routledge, 2025. | Series:
     Routledge research in planning and urban design | Based on the author's
     thesis (doctoral)—University College, London, 2021, under the title:
     A marvellous order : how spatial and economic configurations interact
     to produce agglomeration economies in Greater Manchester. | Includes
     bibliographical references and index.
Identifiers: LCCN 2024036260 (print) | LCCN 2024036261 (ebook) |
     ISBN 9781032384566 (hardback) | ISBN 9781032394961 (paperback) |
     ISBN 9781003349990 (ebook)
Subjects: LCSH: Public spaces—England. | Public spaces—United States. |
     Urbanization—England. | Urbanization—United States.
Classification: LCC HT185 .F76 2025 (print) | LCC HT185 (ebook) |
     DDC 307.760942—dc23/eng/20241014
LC record available at https://lccn.loc.gov/2024036260
LC ebook record available at https://lccn.loc.gov/2024036261

ISBN: 978-1-032-38456-6 (hbk)
ISBN: 978-1-032-39496-1 (pbk)
ISBN: 978-1-003-34999-0 (ebk)

DOI: 10.4324/9781003349990

Typeset in Times New Roman
by Apex CoVantage, LLC

This book is dedicated to my father, Martin Froy, who handled complexity so beautifully in his paintings.

Contents

List of figures and maps	*xi*
List of boxes	*xii*
Acknowledgements	*xiii*
Introduction	1

PART ONE

1	Why focus on urban complexity?	5
2	Theories of complexity	10
3	Parts and wholes: the configuration of urban economies	19
4	Parts and wholes: the configuration of urban space	36
5	Bringing together configurational analysis of economies and space	47

PART TWO

6	How local economies evolve and branch	65
7	How spatial complexity evolves and supports branching economies	80
8	Waterproofing: a case study	98
9	The destruction of urban complexity	105

x *Contents*

PART THREE

10 Cities as systems of systems: where nature fits in 127

11 Rebuilding urban complexity: how can policy makers
    intervene? 139

Index *157*

Figures and maps

3.1	Supply chain matrix for Greater Manchester	23
3.2	Skills-relatedness matrix for Greater Manchester	25
3.3	The position of economic communities in the economic hierarchy of cities	26
3.4	Skills basins in Greater Manchester	27
3.5	The textiles and clothing skills basin	28
4.1	Local spatial integration in Greater Manchester	42
4.2	The configuration of Trafford Industrial Park compared with the city centre	43
4.3	The foreground network overlain on a map of building densities	44
4.4	Sheffield – locating the "heart of the city"?	45
5.1	Bee imagery in Manchester	48
5.2	Commuting flows between Greater Manchester local authorities	53
5.3	Fashion and textiles cluster in Strangeways	56
5.4	Positioning of Strangeways between two spokes of the foreground network	57
5.5	Mapping "fields of collaboration potential" in Altrincham	58
6.1	Historical economic branching in Greater Manchester	67
7.1	The foreground network of the 1850s city	83
7.2	18th-century grid of streets in Ancoats	85
7.3	Land uses in the area around Dale Street, 1886: Annotated GOAD Fire Insurance Plan	87
7.4	Building footprint in the area around Dale Street, 1886	88
7.5	Diverse land uses south of the River Irwell in Broughton, 1952	91
7.6	Change in use of plots between 1850/1890 and 1950	92
8.1	Diversification surfaces in the history of waterproofs	100
9.1	Map of proposed ring roads in 1945	107
9.2	Changes to the Hulme urban fabric from 1942 to 2006	109
9.3	The Junction Pub in Hulme, 1984	109
9.4	Park Hill Flats in Sheffield	111
9.5	The railway viaduct coming through Newcastle quayside	113
9.6	Disurbanism in the heart of Newcastle	114
9.7	A mural on a pumphouse in Broughton, Salford	120
11.1	Three different ways of knitting the urban fabric back together	145

Boxes

Box 1	Gephi – a tool for visualising networks	23
Box 2	Wright Bower: participating in global production chains under the arches	32
Box 3	Stockport and its hat cluster	50
Box 4	Not everybody wants to share their knowledge	54
Box 5	Textiles – a network-based material	68
Box 6	Tottenham Court Road – home to furniture since the 18th century	92
Box 7	Private White V.C. – a brave manufacturer?	100
Box 8	Jay Trim – a firm that has adapted to urban spatial change	115
Box 9	Contamination – when complexity becomes problematic	130
Box 10	Lessons not learnt from "pop-up" developments in Folkestone	147
Box 11	Learning from Hackney in London	149
Box 12	"A protected space" – the Grainger covered market	150

Acknowledgements

First and foremost, I would like to thank my mother, Catherine, who has been a sounding board throughout my writing, providing invaluable guidance again and again on how to order my thoughts and express myself clearly. I am not sure how people write books without such support! At Routledge, I would like to thank Fran Ford, Caroline Church, Meghna Rodborne and Sangeetha Santhanam for seeing the potential of this book and guiding me so well during the whole publishing process.

I am deeply grateful to the people of Greater Manchester – from bag makers and waterproof coat manufacturers to policy makers to industrial archaeologists to artists – who so generously gave their time and knowledge. Specific thanks go to John Holden and Rupert Greenhalgh, who helped me get my research off the ground. I would also like to thank Dr. Jessica Symons and Mark Treacher for hosting me during my fieldwork.

At University College London, I would like to thank Dr. Sam Griffiths and Professors Laura Vaughan, Alan Penn and Neave O'Clery for all their advice during the PhD which was the basis for this book (funded by a grant from the Engineering and Physical Sciences Research Council). I also had the privilege of studying with Professor Bill Hillier – such an inspiring and also considerate academic. Elsewhere, a range of people with an interest in systems and complexity from diverse disciplines have helped to shape my ideas, including Ed Parham (Space Syntax); Professor Elvira Uyarra, University of Manchester; and Professors Howard Davis and Peter Lloyd, who walked with me around Manchester and shared their observations. Thank you to Dr. Nicolas Palominos, for his ongoing research collaboration, and for so willingly volunteering to produce a number of the maps in this book. Jacob Miller, Subik K. Shrestha and Merve Okkali Alsavada also helped with specific research tasks, while Leah Aaron provided useful comments on a draft. I much benefitted from recent conversations with Pete Stringer, Nalin Seneviratne and my sister Amanda Glasspole about Manchester, Sheffield and Folkestone respectively. Jeanne and Isabelle Ruscher helped me to choose between endless possible titles. At the University of Oxford, I would like to thank Dr. Patricia Canelas, Dr. Nigel Mehdi, Dr. Desiree Daniel-Ortmann and Dr. Pedro Marquez from the Sustainable Urban Development team for their support, inspiration and conversations which helped me to get this book over the finishing line – in addition

xiv *Acknowledgements*

to the master's students with whom I have had such interesting discussions about systems theory.

My grandfather, Eric Hodgson, inspired me in part to write this book, due to his role as a factory inspector, his collection of books on industrial archaeology and his great interest in buildings and the morphology of villages.

Thank you finally to my partner, Chris McDonald, for being a constant support during the writing process, and for putting up with my laptop appearing in strange places, including on ironing boards and the edges of swimming pools.

Introduction

This is a book about urban complexity – how it evolves and how it gets destroyed. When we describe a system as complex we mean that there are interdependences between its parts, and the parts come together to create an "emergent whole", which is not explainable by simply looking at the individual parts alone. In the following chapters, we will explore the particular structures of interdependency which underpin cities, where the many different "parts" (such as people, streets, businesses, rivers) interact within an evolving "whole". We will consider how cities are internally "self-supporting" but also open to broader international networks of trade, migration and knowledge exchange. We will look at questions of stability and resilience, and identify how cities lose some of their adaptability when intricate patterns of complexity are disrupted through insensitive top-down planning interventions.

The book has a particular focus on urban *space* (drawing on architectural theories) and *economic capabilities* (drawing on economic geography), but it is also interdisciplinary in scope, drawing in threads from history, anthropology and ecology. We will later consider how cities can better harness the self-organising capacities of the natural environment, while learning lessons about the "lattice economy" which existed in Victorian cities as an important recycler of waste. We will explore the evolution and destruction of complexity in one city – Greater Manchester – in particular, but also compare it with other post-industrial cities: from its neighbour Sheffield, to Newcastle in the northeast of England, and cities further afield such as Detroit, Michigan and New Haven, Connecticut. We will conclude by considering how cities might rebuild urban complexity, following relatively simple rules to recreate the urban integrity and "marvellous order" which is so important to sustainable urban development.

DOI: 10.4324/9781003349990-1

Part One

1 Why focus on urban complexity?

Following a career in local economic policy lasting over three decades, this book has arisen from my desire to move beyond the standard ways of intervening in cities through top-down local economic strategies and spatial master plans. When I started out in this field I lacked a perspective of how urban economies evolve and develop – falling into the frequent policy practice of "muddling through" based on previous policy implementation rather than grounding my actions in in-depth research into what makes cities work [1].[1]

The disconnect between local policies and underlying capabilities

I learnt a great deal from working in international organisations in Brussels and Paris, where I witnessed local economic development policies and practices from around the world. Through roles evaluating European policy in Brussels and later working as a policy analyst at the OECD (Organisation for Economic Cooperation and Development) I had the privilege of visiting many places, from Flanders in Belgium to the Bay of Plenty in New Zealand. I evaluated local, national and international policies that aimed to turn around local economies that had fallen into decline and talked to people about their main local challenges and their visions for creating a better future. On the basis of these visits and conversations, I and my colleagues developed a series of books, policy working papers and international conferences on local economic development and local governance.

However, as my career progressed, I became increasingly uneasy with the advice that we were providing to local policy makers which reinforced widely held, "common sense" beliefs that local economies could be effectively *planned* and managed. It was thought that as long as the governance of a particular place was carried out effectively and a well-designed local economic strategy was in place, then a place would prosper. However, I saw that such policies often had only a marginal effect and that there was a degree of policy inertia, with the same strategies being repeated year after year and in place after place without a full assessment of their impact. In parallel, little attention was being paid to the underlying characteristics and potentials of specific local economies. A dearth of local data meant that people were only dimly aware of the capabilities embedded in their local industrial structure and the spatial characteristics of their local built environment. While

DOI: 10.4324/9781003349990-3

6 Rebuilding Urban Complexity

long-term strategies abounded, many local institutions were in fact mostly working in reactive mode, dealing with problems as they arose. As an example, local educational and employment officials were often "fire-fighting" to solve immediate skills gaps and unfilled vacancies [2], rather than trying to understand the skills and capabilities that were already embedded in the workforce, offering latent potentials for future economic development. In addition, there was little acknowledgement among local policy makers that the *built environment* of cities might have a role in shaping economic prosperity. Very often the built environment was seen as a background to the real drama, which was the plethora of local institutions trying to make a difference to local poverty and productivity. I started to question how and to what extent architectural form might be influencing economic outcomes.

Coming to appreciate the importance of self-organisation in local development

At a conference in Madrid, I was introduced to the works of Jane Jacobs, who offered a very different view– far from the policy delivery interface – of how urban economies operate. Like many others, I became captivated by her books *The Death and Life of Great American Cities* and *The Economy of Cities*, which highlighted the fact that urban economies to some extent evolve on their own, having their own powers of self-organisation. Jacobs revealed to me the "organised complexity" of cities and the 'intricate social and economic order under the seeming disorder of cities' [see, e.g., 3, p. 21–22]. She firmly rooted her accounts of urban life and urban evolution in the built environment – describing in detail the urban spatial structures that underlie how people move through the city, encounter each other and develop thriving forms of social and economic life. She highlighted the branching processes through which current economic practices diversify into new forms of innovation, again through bottom-up forms of self-organisation. Seen through this lens, economic transformation is not something that you impose or manage "top down". It is rather something that emerges through networks of interdependency and through multiscale forms of complexity. Urban economies can be harnessed and nurtured by policy makers who recognise the latent potentials in local emergent systems – but they are very difficult to design via a 100-page local economic development plan. Jacobs emphasised that understanding the *spatial* dimensions of a city – and the network structures of public space which underpin them – was crucial to understanding how cities function. And she reiterated that cities were very difficult to simply spatially "design" through a top-down master plan, without fully understanding the complex multiscale interdependences which allow spatial networks in cities to function.

My reading of Jane Jacobs prompted me to start reading more broadly about processes of complexity and emergence in both economic geography and architecture. These ideas revived an academic interest, which had begun when I was studying anthropology back in the 1990s, in the underlying systems (often material) which shape our everyday lives. To deepen my knowledge of the spatial qualities of local built environments, I spent more than ten years exploring Bill Hillier's

Why focus on urban complexity? 7

ideas of "space syntax" at the Bartlett School of Architecture in London. Hillier and his colleagues had transformed Jacobs' observations about the importance of the anatomy of streets into a way of understanding cities as products of complex and multiscale spatial configurations. I started to understand that the built environment provides a set of "affordances" [4] which support social and economic interactions in cities. I realised that a combination of network analysis and also a "deep description" of actual processes, down to a fine grain, would be necessary to more fully understand how a city might be operating. Whilst still at the OECD, I was introduced to Frank Neffke of the Harvard Growth Lab, who was working with a team of scientists to explore interdependences between economic capabilities at both the national and regional scales. I noted that despite being immersed in urban data science, many of the theorists I encountered at the Bartlett and the Harvard Growth Lab continued to quote the work of Jane Jacobs, building on her patient observation of actual urban processes and providing evidence for her thinking through mathematical techniques (in particular graph theory).

Focusing on Greater Manchester and other post-industrial cities

I looked for a city to explore in detail, where I could apply my newfound economic and spatial theoretical understandings in practice. The last international forum meeting which I helped to organise for the OECD was in Greater Manchester. We discussed many different aspects of local governance and international practice at this meeting. However, my new grounding in network science allowed me to have very different conversations with local policy makers and academics about Greater Manchester's spatial and economic complexity. The national government also showed strong interest in the city, which it saw as sitting at the heart of policies to rebalance the relatively London-centric UK economy, through the 'Northern Powerhouse', and later 'Levelling up' [5]. I decided to delve deep into the urban economic and spatial complexity of the city via a PhD thesis, based at the Bartlett School of Architecture.

This book is partly based on this PhD research. It explores the spatial and economic networks and structures that have underpinned the economy of Greater Manchester from the 18th century to today. However, it is not just about this one city. My explorations revealed to me the spatial and economic complexities which underpin any great city – and particularly cities which have had an industrial past and are now trying to reinvent themselves with new service-based and knowledge-based economies. It is therefore 'told with the specific and particularity of the city I know best, but told with a thought to the other places that have written broadly similar histories of their own' [6, p. xvi], so following the example set by Douglas Rae in his explorations of New Haven, Connecticut, in the United States.

I also draw on research in other post-industrial cities to shed new light on how city economies more broadly evolve, prosper, decline and then reinvent themselves. Post-industrial cities such as Sheffield and Newcastle have also seen a gradual evolution of complexity, and then experienced an important and sudden loss

8 *Rebuilding Urban Complexity*

of their finely integrated network infrastructures due to insensitive urban planning and transport interventions in the mid-20th century. This *destruction* of complexity is as important to understand as the processes which lead to the emergence of complexity. Like Douglas Rae, we explore the 'brave strokes of physical renewal, their social costs, and some of their unintended consequences' [6. p. xv].

In this book, the reader will be introduced to complexity and systems theory and a set of architectural ideas about how urban space can work to produce lively and innovative cities. The analysis uses mathematical modelling and graph theory to reveal urban potentials. Urban data is hard to come by for mapping local supply chains and labour flows between industries. Where local data is not available, models are used to predict likely economic interdependencies in a city and the latent potentials present in economic and spatial structures. However, reading this book does not require a grasp of mathematics or econometrics. Network analysis is supplemented with qualitative material from historic archives, interviews and fieldwork to tell a story indivisible from its context.

As the book explores the diverse economic capabilities which cities host, the focus is as much on their material capabilities as their social or cognitive ones. As human capabilities and knowledge evolve, they become incorporated and embedded in technologies, products and materials. This book foregrounds the infrastructures and material artefacts that all too easily slip from our consciousness as we try to understand how cities work [see 7].

Finally, the book focuses both on the past – the processes of historical evolution that are so important to urban complexity – and the future, seeking to understand how cities can reposition themselves in a global economy which is requiring a transition to less carbon-intensive modes of production and economic practices that are less harmful to both local and global biodiversity. It looks at processes of urban resilience but also *reinvention*. In its conclusions, the book therefore circles back to the role of local economic, architectural and planning policies in rebuilding urban complexity. I will reveal how local policy makers can support both the resilience and reinvention of cities, without having to impose top-down forms of order which at best "jar with" and at worst destroy the evolved order of things.

The book has three parts, focusing in turn on three dimensions of urban complexity: 1) parts and whole structures, 2) evolution and branching, and 3) cities as "systems of systems". In each part, we will move between focusing on the complexity of the built environment and the complexity of economic systems, and then consider how the two perspectives can come together. As noted earlier, the book concludes by considering policy implications and how policy makers can best respond to, and rebuild, systems of emergent self-organisation in post-industrial cities.

Note

1 As a young consultant in the 1990s I evaluated and advised on employment and training schemes in Manchester and Liverpool, and later joined the economic development and housing teams for Reading Borough Council.

References

1. Lindblom, C., The science of 'muddling through'. In *Classic readings in urban planning*. 2018, London: Routledge. p. 31–40.
2. Froy, F., Global policy developments towards industrial policy and skills: skills for competitiveness and growth. *Oxford Review of Economic Policy*, 2013. **29**(2): p. 344–360.
3. Jacobs, J., *The death and life of great American cities [1993 edition]*. 1961, New York: The Modern Library.
4. Gibson, J.J., *The ecological approach to visual perception*. 1979, Boston, MA: Houghton Mifflin.
5. Tomaney, J. and A. Pike, Levelling up? *The Political Quarterly*, 2020. **91**.
6. Rae, D.W., *City: urbanism and its end*. 2005, New Haven: Yale University Press.
7. Star, S.L., The ethnography of infrastructure. *American Behavioral Scientist*, 1999. **43**(3): p. 377–391.

2 Theories of complexity

This book is underpinned by "complex systems theory" – so we will start by exploring what it is, and how it might help us to understand cities in the abstract. Systems thinking has emerged in a relatively fragmented way across many different disciplines, and indeed elements of it can be traced back to very early philosophies and religions. Much systems thinking comes relatively intuitively to us – we are used to acknowledging "part-whole" relationships and the dynamic changes and unpredictability which govern our everyday lives. Negotiating daily with the internet and its associated webs of social media, we are increasingly aware of networks and their power to shape us.

An understanding of the importance of the networked and complex aspects of life began to penetrate scientific and academic discourse in the West from the beginning of the 20th century. A broad trajectory of systems thinking can be traced from the work of the evolutionary biologist Ludwig von Bertalanffy at this time through to *cybernetic* theory from the 1940s onwards (the work of theorists such as Norbert Wiener, Margaret Mead and Gregory Bateson), and then the emergence of theories of chaos and *complex adaptive systems* in the 1960s and 1970s (involving theorists such as Nicolis and Prigogine, and Stuart Kauffman) [1]. While the emergence of cybernetics at the end of the Second World War brought with it a concern with management and control, the latter theorists reversed this trend. They emphasized that many of the most successful systems operate at the "edge of chaos", incorporating bottom-up dynamic processes of change which are largely unpredictable, with every situation having multiple different "adjacent possibles".

Systems and complexity theories may vary in their approach and their key areas of focus, but they share a common basic framework or "ontology". In the 1960s, Von Bertalanffy published a general theory of systems [2], describing the *isomorphism* which exists between systems, with common systemic attributes occurring in the organisation of very different phenomena. This means that we can use a common language to describe systems properties and a common set of techniques for understanding them. As a result, systems thinkers tend to work across disciplines – the work of Stuart Kauffman, for example, ranges from cell biology to anthropology to economics. Gregory Bateson applied his communication theories at multiple scales, from octopus communication to human communication systems in the heat of war. The Santa Fe Institute, which is a world-class school in

DOI: 10.4324/9781003349990-4

Theories of complexity   11

complexity science, incorporates a mixed group of physicists, biologists, economists and political scientists.

A core principle within systems theories is that the world cannot be reductively
understood as being composed of simple elements or "parts" which have predictable and simple linear relations with each other. Rather, the world is characterised
by multiple interdependences and network relationships through which elements
often combine to form wholes that are "more than the sum of their parts". Donella
Meadows of the Massachusetts Institute of Technology described how 'a system
is a set of things – people, cells, molecules, or whatever – interconnected in such a
way that they produce their own pattern of behavior over time' [3, p. 2]. Complexity systems thinkers point to the uncertainties that exist regarding the outcome of
such interactions, but concur that systems often have a tendency to fall into certain
structures, known as "attractors", which can remain stable for long periods of time.
Such attractors can be both physical and social. The anthropologist Mary Douglas,
for example, gradually brought systems thinking into her analysis of social and
political systems as part of an effort to understand the basic orderings which allow
human societies to cohere (as opposed to simply falling apart). She identified the
feedback loops that allow societies to be maintained, while also pointing to four
principal possible attractor states which societies were likely to fall into (isolates,
individualism, hierarchy and self-similar enclaves) [4].

As systems thinking has progressed, such sociopolitical components of systems have moved to centre stage, with critical systems thinkers exploring, for
example, the importance of power relations in driving different systems. At the
same time, theorists such as Donnella Meadows emphasise that systems are not
just "out there", in the real world, but also embedded in our thinking in the form
of paradigms, mental systems and worldviews. In the field of cybernetics, Margaret Mead identified the importance of developing a 'second order cybernetics'
which identified that theorists themselves are part of the system, forcing a degree
of self-reflection and reflexivity.

Key concepts and principles in complexity and systems thinking

In the remainder of this chapter, we explore eight common principles which will
be of importance as we start to explore concrete, city-specific findings in the rest
of the book.

1) *Complex systems are relational*

The first of these principles is interrelatedness. Complexity and systems thinkers
emphasize the *interdependencies* between things, seeing phenomena as *relational*.
An example is the ecosystem in biology, where plants and animals are not just seen
as individual species but rather as species that come together to form enduring relationships in a particular environmental niche. Systems theorists therefore have a
strong interest in "networks" and how they function, drawing on the mathematical
theory of graphs to better understand the interrelationships between the different

12   *Rebuilding Urban Complexity*

parts of a system (with the parts being "nodes" and the interlinkages between them being "edges").

2) *Not all relationships are equal*

Because of their interest in interrelatedness, systems thinkers are also concerned with "topology", which is defined by the Oxford English Dictionary as 'the way in which constituent parts are interrelated and arranged' or more specifically, as those relational and networked properties 'which are independent of size and shape and are unchanged by any deformation'.[1] When networks are analysed in systems theory there is a recognition that not all interrelationships are equal. Thinking *topologically* suggests that certain parts of a system may be privileged in terms of how they relate to each other, with some parts of the system being "shallower" or "deeper" than all the other parts. Depending on where different nodes are positioned, they have a more or less privileged relationship in respect to the rest of the system. Some of the "edges" or connections between nodes have more "traffic" on them, and thus have a higher "edge-weight". Network theorists such as Barabási [5] also focus on how networks in real-world systems (such as the internet) are often highly skewed, with a small number of nodes having a relatively high number of strong connections, while the rest of the network is composed of "weak links".

3) *The whole is bigger than the sum of its parts*

The interrelationships between different parts of a system operate to ensure that the system produces something that is "more than the sum of its parts". This can occur in a social and economic system, such as a city, or a biological ecosystem, such as a coral reef. Donna Meadows helped people to identify the existence of systems through a sequence which starts with identifying "parts", and then asks whether these parts together produce an effect that is different from the effect of each part on its own, and whether this behaviour to some extent persists over time [3]. When it comes to cities, Jane Jacobs famously pointed to the myriad interdependencies which come together to produce systems of "organised complexity". In *The Death and Life of American Cities*, she harnessed the latest ideas in complexity theory, drawing on Warren Weaver's classification of relationships into: 1) simple bilateral relationships between one factor and another (which produce linear relationships that can be more easily analysed through science); 2) disorganised complexity (which can best be analysed through statistics); and 3) organised complexity, which for Jacobs involved 'dealing simultaneously with a sizeable number of factors which are interrelated into an organic whole' [6, p. 563].

The economic geographers Ron Martin and Peter Sunley [7] shed more light on this topic by describing how the relationship between parts and wholes is what characterises systems as being *complex* as opposed to merely *complicated*. They argue that "a system is complex when it comprises non-linear interactions between its parts, such that an understanding of the system is not possible through a simple reduction to its component elements" (p. 577). A similar distinction was made at

Theories of complexity 13

an event organised by the OECD Global Science Forum in 2009, which explored how complexity science could be of greater use to policy makers. In the words of the OECD report:

An example of a *complicated* system is an automobile, composed of thousands of parts whose interactions obey precise, simple, known and unchanging cause-and-effect rules. The complicated car can be well understood using normal engineering analyses. An ensemble of cars travelling down a highway, by contrast, is a *complex* system. Drivers interact and mutually adjust their behaviours based on diverse factors such as perceptions, expectations, habits, even emotions.

[8, p. 2]

It is not just the nature of parts which affects how they come together in complex systems as wholes – it is also the structure of the relationships between them. Again, the *topology* of systems plays an important role – as Glückler and Doreian point out, 'the structure of networks affects collective outcomes' [9, p. 1124].

4) *Structure arises through self-organisation*

Another important systems characteristic is *self-organisation*. Parts produce wholes through an emergent process, whereby a myriad of interactions between the parts produces order "autopoetically". Systems therefore do not need to be "planned" or to have external influence in order to change – all that is necessary is an underlying source of energy for systems structures to emerge from such interactions between their parts. Complexity scientists refer, for example, to the emergence of macro-structures from 'microevents and behaviours' [10]. Often local rules are sufficient for order to emerge – an example being how the local rule that a starling follows the flight of its neighbour ultimately leads to the beautiful group formations known as murmurations.

Another key way in which self-organisation occurs is through feedback loops. Feedback loops can be sustaining of the current system (an example being our own bodies, where our hormonal system produces a degree of homeostasis – in terms of stable temperature, water, sugar and salt levels). However, they can also be amplifying. Mary Douglas, for example, analysed the role of feedback loops in situations of conflict and extremism – something that Franz Fanon also famously pointed to when he identified that violence (including the dehumanising violence associated with colonialism) begets violence [11]. Such processes are also encapsulated in a biblical reference from St Matthew: 'For to everyone who has, more will be given' [see 3].

5) *Systems have hierarchy*

Another important dimension to systems is that the part-whole structure has *hierarchy*. When parts come together into "wholes", generally intermediate layers of

14   *Rebuilding Urban Complexity*

organisation can be identified. The political scientist and economist Herbert Simon [12] identified hierarchical "community structure" as being an important, defining characteristic of complex systems. Systems are generally multiscale, with each level in the hierarchy having its own emergent properties and causal capacities. In the context of social systems, this might equate to pupils at a school forming different friendship groups in the playground which can then be mapped through network analysis. Depending on the scale at which theorists look at a system, they will find different effects. It is important, therefore, to always pay attention to the "scale of resolution". The significance of this quickly becomes apparent when exploring the economic diversity which exists in cities. What at first glance seems like the "same" – i.e. a single economic sector – is quickly revealed as itself constituted by a myriad of different parts. The cyberneticist Gregory Bateson identified that 'what can be studied is always a relationship or an infinite regress of relationships. Never a "thing"' [13, p. 246]. This also offers insight into the fractal nature of many spatial and economic realities [14, 15].

6) *Most systems are open*

In order to persist, self-organising systems need to have a degree of internal coherence – incorporating enough "closed circuits" to be self-sustaining [16, p. 265]. However, most systems are also open – that is, they form wider interdependencies and relationships. It is clear that cities, for example, are embedded in multiple national and international relationships which are key to their economic prosperity. The architecture theorist Stephen Read and colleagues suggest, 'what we begin to understand about urban structures is that where they work, they do so by opening the 'outside' world to us 'in' our local places, and by bringing the potentials of the world to hand' [17, p. 16]. Because systems are open, a degree of judgement must be made when it comes to how one sets the "boundary" of the system that one is looking at. If the boundary is too tight, this can lead to the neglect of important factors which are actually crucial to the reproduction of the system itself. Ecologist James J. Kay describes, for example, how when the Aswan Dam was constructed on the Nile delta to produce hydroelectric power, the engineers failed to consider how this would interrupt the annual flooding of the Nile along its course [1]. This loss of flooding meant that nutrients were lost to downstream agriculture, which in turn meant that new forms of energy had to be sourced to produce fertilisers, negating the energy benefits produced by the dam. By leaving agriculture and downstream impacts out of the model, the whole rationale of the project was undermined.

7) *Systems are mostly only partially ordered*

The complexity systems theorists of the 1960s and 1970s pointed out that while the interdependences which form systems have *structure*, systems also very often host a high degree of randomness – which is what ensures that effective systems are often operating "at the edge of chaos" (being not too rigid and not too fragmented).

Theories of complexity 15

We will see in the analysis of city networks throughout this book that structuring tendences only provide a weak restriction on a largely random set of processes. The degree to which there are "necessary relations" between things is very small. The built environment, for example, is not seen to determine behaviours in any sense – it rather creates 'a field of possibilities and restrictions' [18, p. 348]. Analysts of economic complexity also recognise that there is generally a large amount of "noise" when attempts are made to uncover the most likely interrelationships between industries.

8) *Complex systems are dynamic and evolving*

Finally, systems are rarely static – they evolve. Systems thinkers often talk about the notion of "emergence" – as we saw previously, this term is used to understand processes of self-organisation which allow wholes to emerge from their parts. However, the term "emergence" is also used to describe how systems slowly and incrementally evolve over time, powered by their own internal possibilities while also being influenced by outside processes. This is often a "branching" process, which will be familiar to all who understand Darwinian evolution. Complexity theorists such as Stuart Kauffman point to the fact that branching is a more general property of systems, including economies. Jane Jacobs likewise referred to a general process by which 'differentiations become generalities from which further differentiations then emerge' [19, p. 17].

In order to imagine possible future evolution, systems theorists often talk about 'latent potentials' [3]. These are potentials in the current system which make it likely that the system will evolve in a certain way. As we have already seen, theorists such as Stuart Kauffman also identify how systems have a set of "adjacent possibles" – likely new states that might come to exist based on the structures and "preadaptations" present within the current system [20–22]. The idea that what comes next is shaped by myriad interdependent conditions that exist in the present is also encapsulated in the Buddhist idea of "contingent arising" (for an exploration of the links between systems thinking and Buddhism see [23]). Systems theorists advise people who propose to intervene in systems to pay attention to such latent potentials so that they can "go with the flow". As an example in the field of ecology, González discusses 'restoration plans that work in synergy with the tendencies that arise from the systems' structure' [24, p. 8]. Likewise, when considering how complexity theories can best be harnessed by policy makers, Cesar Hidalgo talks about 'paths that may be easier to climb' [25, p. 13]. DeLanda [26] identifies such 'virtual' properties of existing systems to be a key dimension of reality, constituted by (as yet) unactualised tendencies. Read et al. [17, p. 1] also discuss 'metropolitan landscapes of actuality and potentiality'.

More "resilient" systems are felt to evolve in ways that tolerate and adapt to outside shocks, while preserving their internal coherence through "structure-preserving transformations". The term "resilience" was first used in the systemic sense in ecological research and literature about vulnerability to natural disasters, before becoming widely used in economic geography in recent decades [27–29]. There

16 *Rebuilding Urban Complexity*

are different schools of thought as to whether resilience represents the possibility of "returning to a previously existing equilibrium" or rather a process of continuous evolution and adaptation (with this book arguing more for the latter).

An important aspect of systems resilience is "redundancy", wherein resilient systems have a "broad playing field" incorporating duplication and seeming inefficiency, which means that should one part of the system fail, it is easy to reroute connections and replenish the overall system [3]. Internal feedback loops also often maintain systems in a stable state despite external pressures – at least up to a certain point. James J. Kay, for example, points to the fact that the pH level of lakes often remains stable over relatively long periods of time, despite being bombarded with acid rain. A series of natural feedback loops keep lakes within a certain pH range. Nevertheless, when a particular level of acidity is exceeded, there is a "tipping point" which means that the compensatory effect of feedback loops is no longer sufficient, and the lake suddenly becomes acidified [1]. Systems can move to a new "attractor" state – another relatively stable system which can be radically different from what preceded it. Tipping points are everywhere, and are particularly worrying in the context of climate change. Ecologists point to the fact that the Earth has a series of buffers and feedback loops which are constraining the rapidity and impact of human-induced temperature rises. Nevertheless, once certain tipping points are reached (due, for example to the melting of the Arctic ice cap, or the collapse of the Gulf Stream), changes could be much more dramatic.

Where does materiality and infrastructure fit in?

All of these principles come into play as we explore complex systems in specific cities in this book. As already highlighted in the introduction, we will look in particular at *materially embedded* systems – such as those associated with buildings, streets, tools and fabrics. It is perhaps worth saying a bit more, then, about how materiality has become a preoccupation for systems thinkers in recent decades. During the 1980s, "actor-network theory" emerged in France which emphasised that social systems have a strong material component, and indeed that material elements within systems can have agency in a similar way to human elements. Bruno Latour, for example, talks about the 'missing masses' [30], describing the material elements that are overlooked in accounts of social behaviour. The agency connected to material objects has become the topic of theories of "assemblage" (associated with actor-network theory but also with the work of the French theorists Gilles Deleuze and Félix Guattari, and later the Mexican-American theorist Manuel DeLanda) which explore the networked interdependences that exist between both human and nonhuman actors.

In this book, the material world is conceived of as embodying latent potentials and underlying capabilities which are important to how cities both work and evolve, and it therefore aligns with actor-network theory. Departing from this theory, however, it follows space syntax thinking in suggesting that the built environment can take on a special role in systems, not just as one material actor alongside all others, but rather as an underlying infrastructure which makes other social and

Theories of complexity 17

material interdependences possible. Our streets and buildings make a special contribution to what is sometimes known as "systems-level agency". Street systems "assemble" and generate relations [31]. Following the thinking of ecologists such as Paul Jepson, networks of streets and public spaces in cities are also shown to provide the basis for a broader systems *integrity* (linking parts into wholes) [32]. In ecological systems, coral creates a material substrate for the rich diversity of coral reefs. Likewise, in the built environment, streets and buildings are seen to provide an important underpinning for both economic and human diversity – acting, in the words of DeLanda, as "intercalary" elements which are "quasi-causal" in that they only ever partially order broader interdependences [16, 26]. They bring together the other forms of diversity that exist in cities, in a way which 'meshes multiplicities by their differences' (26, p. 103). When visiting neighbourhoods that are buzzing with life – such as the Rue de Commerce area, where I used to live in the 15th arrondissement of Paris, or the Jericho area of Oxford – it is apparent that there are multiple overlapping systems operating in synergy, with the built environment hosting many different systems (commercial, civic, family, religious, ecological) which combine to create a feeling of successful and enduring urbanity.

Armed with these key systems concepts, we can now start to explore how they can be applied to our key focus here: post-industrial cities. We will focus on two interacting systems in particular – economic systems and spatial systems – but also on socioecological systems and the potentials created by nature-based infrastructures. While we can isolate systems in cities to better understand them, it is impossible to fully separate them from the rich assortment of other systems in which they are embedded. Cities are always "systems of systems".

Note

1 Oxford English Dictionary, Online Edition. www.OED.com. Accessed [07/06/2024].

References

1. Kay, J.J., An introduction to systems thinking. In *The ecosystem approach: complexity, uncertainty, and managing for sustainability*, D. Waltner-Toews, J.J. Kay and N.E. Lister, Editors. Chichester: Columbia University Press, p. 3–13.
2. von Bertalanffy, L., *General system theory: foundations, development, applications.* 1968, New York: G. Braziller.
3. Meadows, D.H., *Thinking in systems: a primer.* 2008, White River Junction, VT: Chelsea Green Publishing.
4. 6, P. and Richards, P., *Mary Douglas: understanding social thought and conflict.* 2022, New York, Oxford: Berghahn Books.
5. Barabási, A.-L., *Linked: the new science of networks.* 2003, London: Plume Books.
6. Jacobs, J., *The death and life of great American cities [1993 edition].* 1961, New York: The Modern Library.
7. Martin, R. and P. Sunley, Complexity thinking and evolutionary economic geography. *Journal of Economic Geography*, 2007. **7**(5): p. 573–601.
8. OECD Global Science Forum, *Applications of complexity science for public policy: new tools for finding unanticipated consequences and unrealised opportunities.* 2009, Paris: OECD Publishing.

18   *Rebuilding Urban Complexity*

9. Glückler, J. and P. Doreian, Editorial: social network analysis and economic geography – positional, evolutionary and multi-level approaches. *Journal of Economic Geography*, 2016. **16**: p. 1123–1134.
10. Arthur, W.B., Self-reinforcing mechanisms in economics. In *The economy as an evolving complex system*, P. W. Anderson, Editor. 1988, Boca Raton: CRC Press, p. 9–31.
11. Fanon, F., *The wretched of the earth*. 1963, New York: Grove Press.
12. Simon, H.A., The architecture of complexity. *Proceedings of the American Philosophical Society*, 1962. **106**: p. 467–482.
13. Bateson, G., *Steps to an ecology of mind*. 1972, San Francisco: Chandler Publishing Company.
14. Callahan, G. and S. Ikeda, *Jane Jacobs, the anti-planner*. Mises Daily Articles. 2003, F.A.P. Mises Institute, Austrian Economics. https://mises.org/mises-daily/jane-jacobs-anti-planner. Accessed [02/10/2024].
15. Batty, M., P. Longley, and S. Fotheringham, Urban growth and form: scaling, fractal geometry, and diffusion-limited aggregation. *Environment and Planning A*, 1989. **21**(11): p. 1447–1472.
16. DeLanda, M., *A thousand years of nonlinear history*. 1992, New York: Zone Books.
17. Read, S., et al., *Constructing metropolitan landscapes of actuality and potentiality*. 6th International Space Syntax Symposium. 2007, Istanbul.
18. de Holanda, F., Sociological architecture: a particular way of looking at places. *The Journal of Space Syntax*, 2010. **1**(2): p. 355.
19. Jacobs, J., *The nature of economies*. 2000, New York: Modern Library.
20. Cooke, P., Introduction: complex systems integration, 'emergence' and policy modularisation. In *Reframing regional development: evolution, innovation and transition*, P. Cooke, Editor. 2013, Oxford: Routledge.
21. Kauffman, S.A., *Reinventing the sacred: a new view of science, reason and religion*. 2008, New York: Basic Books.
22. Hausman, R. and C.A. Hidalgo, *The atlas of economic complexity: mapping paths to prosperity*. 2014, Cambridge, MA: The MIT Press.
23. Theise, N., *Notes on complexity: a scientific theory of connection, consciousness, and being*. 2023, New York: Spiegel & Grau.
24. González, C., Evolution of the concept of ecological integrity and its study through networks. *Ecological Modelling*, 2023. **476**: p. 110224.
25. Hidalgo, C.A., The policy implications of economic complexity. *Research Policy*, 2023. **52**(9): p. 104863.
26. DeLanda, M., *Intensive science and virtual philosophy*. 2002, London and New York: Continuum.
27. Martin, R. and P. Sunley, On the notion of regional economic resilience: conceptualization and explanation. *Journal of Economic Geography*, 2015. **15**(1): p. 1–42.
28. Kitsos, A., A. Carrascal-Incera, and R. Ortega-Argilés, The role of embeddedness on regional economic resilience: evidence from the UK. *Sustainability*, 2019. **11**(14): p. 3800.
29. Holling, C.S., Resilience and stability of ecological systems. *Annual Review of Ecology and Systematics*, 1973. **4**(1): p. 1–23.
30. Latour, B., Where are the missing masses? The sociology of a few mundane artefacts. In *Shaping technology-building society: studies in sociotechnical change*, B. Wiebe and J. Law, Editors. 1992, Cambridge, MA: MIT Press. p. 225–259.
31. Griffiths, S., Manufacturing innovation as spatial culture: Sheffield's cutlery circa 1750–1900. In *Cities and creativity from the renaissance to the present*, I.V. Damme, B. De Munck, and A. Miles, Editors. 2018, New York: Routledge Advances in Urban History. p. 127–153.
32. Jepson, P.R., To capitalise on the decade of ecosystem restoration, we need institutional redesign to empower advances in restoration ecology and rewilding. *People and Nature*, 2022. **4**(6): p. 1404–1413.

3 Parts and wholes

The configuration of urban economies

We start our exploration into urban complexity by looking at "part-whole" relationships – first economic and then spatial. In each of the following two chapters we begin by considering how these relationships are understood and theorised in the academic literature, before going on to explore the complex networked relationships which exist in Greater Manchester and in other post-industrial cities.

Part-whole understandings in economics

While the majority of economists perhaps would not conceive of themselves as "system thinkers", systems principles have nevertheless been evident in this discipline since its origins. One such principle is self-organisation. Back in the 18th century, the economist and philosopher Adam Smith described in detail how the many economic actions which we make day to day lead to an *emergent* market economy which satisfies many (if not all) of our human needs. This is the now famous "invisible hand". There is today a basic assumption amongst economists (at least in liberal democracies) that economies do not need to be planned – and indeed that it would be difficult for human beings to intervene and plan economies given the number of decisions which would need to be controlled and steered [1].

Another system principle which is evident in how economists conceptualise cities in particular is "part-whole" relationships. Economists often characterise cities as examples of economic "agglomeration", noting how the coming together of different economic activities in proximity produces a positive overall effect. In the 19th century one of the founders of neoclassical economics, Alfred Marshall, observed and described the so-called agglomeration effects which existed in the urban industrial centre of Sheffield, and in Manchester and its surrounding towns in Lancashire [2]. He noted the local interlinkages which helped particular clusters (here cutlery making and textiles, respectively) to persist over long periods. He wrote, '[T]he broadest, and in some respects most efficient forms of constructive cooperation are seen in a great industrial district where numerous specialized branches of industry have been welded almost automatically into an organic whole' [3, p. 599]. The interlinkages which Marshall noted included supply chain relationships, labour sharing, and knowledge spill-overs – later known as "sharing, matching, and learning" mechanisms [4]. These economic phenomena are

DOI: 10.4324/9781003349990-5

20   *Rebuilding Urban Complexity*

also often described as agglomeration "externalities", or local impacts that go outside individual firms to affect the broader set of economic activities in a particular place. Marshall famously suggested, for example, that knowledge was transmitted from generation to generation in such clusters of activity through 'being in the air'.

As ideas about agglomeration externalities developed, there was also an interest in "cumulative causation" (or positive feedback loops), with the impact of this systemic effect on local and regional development being explored from the 1940s by the economists Gunnar Myrdal and Nicolas Kaldor. Nevertheless, such observation-based thinking became unfashionable in mainstream "neoclassical" economics, which was increasingly mathematical in its methodologies from the 1950s onwards. Neoclassical economists predicted that markets would always return to perfect equilibrium in time, with balancing feedback loops (such as rising wages in places of labour scarcity, and rising returns to capital when capital is scarce) eventually "ironing out" any changes which occur in particular places. It was only when the American economist Paul Krugman and others began to model agglomeration effects mathematically in the 1990s that theories of agglomeration and an appreciation for cumulative feedback loops came again to dominate economic understanding of cities [5, 6].

Paul Krugman formed part of a broader new orientation towards complexity in economics and economic geography in the 1980s and 1990s. While Krugman and theorists such as Brian Arthur used formal mathematical modelling of complexity to better understand local path-dependent processes, others have explored complexity in more qualitative ways, incorporating a focus on human agency and power relationships in their analyses [7]. The branch of economic geography most akin to my own thinking is "evolutionary economic geography", which has also increasingly explored the ways that cities operate as "complex evolving systems" [7].

The topological structure of diverse local economies

The increasing influence of complex systems thinking in economic geography has resulted in a preoccupation with "relational" systems, based on networked understandings of how economies operate across space [8, 9]. While analysis of economic agglomeration often focuses on single industry clusters (an example being Marshall's interest in cutlery in one place and textiles in another), there has been an increasing recognition that agglomeration mechanisms operate across diverse sectors in cities. The work of Jane Jacobs on urban economic diversity has led to economists now distinguishing between specialisation externalities (rooted in Marshall's thought) and urbanisation externalities ("Jacobs-externalities"), based on economic variety. It is increasingly understood that this economic diversity is not random but rather intricately ordered – with the interdependences which exist across economic sectors having a fine-grained *topological* structure. Cross-sector industrial synergies are found to occur along "tight paths" [10, p. 282], with a few industry sectors showing strong interdependencies, and a long tail of more weakly interrelated industries.

Parts and wholes: the configuration of urban economies 21

This phenomenon has been explored through several different international schools. In the Netherlands and elsewhere in Europe, Ron Boschma and colleagues have established a research focus on "related variety" since the early 2000s [11, 12], whereas across the Atlantic, the Harvard Growth Lab has preferred to focus on what its researchers term "industry relatedness". The latter group takes a more mathematical approach, using graph theory to analyse sets of data which reveal common patterns of interdependence between industries, as opposed to drawing on industrial classifications to understand how industries might be related [see 13, 14]. While much of this analysis takes place at a national scale, various theorists have focused on industry relatedness at the urban and regional scale [see e.g. 15–19]. Through this work, the "sharing, matching and learning" mechanisms which have been explored since Marshall's time can now be explored topologically – revealing the particular configuration of these relationships in local places, and the structured ways in which urban economies "bring their parts together as wholes". Proximity to related industries has been found to confer economic benefits to firms, including employment growth [10, 14, 19], and entrepreneurship survival rates [20], as well as greater resilience after recessions [21, 22].

An example of industry relatedness research which has been influential on my own investigations in Manchester is that by Frank Neffke, formerly based at Harvard Growth Lab and now based at the Complexity Science Hub in Vienna. Neffke has particularly examined "skills-relatedness" – the configurational structure of skills-sharing between industries. To build his relatedness models, Neffke analyses the labour switches or "jumps" which people make between industries during their careers. A higher-than-average number of such jumps between two industries suggests that these industries are particularly skills-related, embodying similar transferable skills, even if each industry uses these skills in a different way. Given the scarcity of data on labour flows at a local scale, Neffke gathers data at a national scale and then models how these relationships might manifest in particular cities given the sets of concentrated local industries that they host. A comparison of skills-relatedness in Amsterdam and Rotterdam, for example, revealed that Amsterdam hosted an important network of potentially interconnected financial and creative industries, while Rotterdam (home to Europe's largest port) saw a greater number of interdependences between logistics and transportation sectors [23]. Otto and Weyh [24] similarly explored the different skills-relatedness patterns which existed around automobile industries in two urban areas in Eastern Germany, revealing how some local industries benefited from these relationships more than others.

Neffke's analysis of interconnected urban labour pools goes some way to revealing what Cesar Hidalgo called a 'collective intelligence' in cities – interrelated capabilities which have emerged in local "constellations" that are difficult to move elsewhere, reminiscent of the letters that form a network of words on a Scrabble board [25]. As previously mentioned, Hidalgo is keen to point out that this collective intelligence is embedded in materials and technologies as much as in individual brains. By focusing on embedded capabilities in industries, as opposed to just the cognitive skills possessed by school-leavers and job seekers, industry relatedness analysis is particularly adept at revealing the broader, materially-embedded knowledge which cities host.

22 *Rebuilding Urban Complexity*

Applying relatedness models to Manchester

In the remainder of this chapter, I summarise my own research into industry relatedness in Greater Manchester. This will in places be necessarily quite detailed and technical, but the broader implications for understanding urban complexity will become evident.

I first looked at the importance of such interdependences for English cities in general. This required applying relatedness models drawn from national and international datasets – particularly as the UK does not routinely collect longitudinal data on labour flows at sufficient scale.[1] Supply chain data was easier to find on the basis of input and output tables, but only at the national level, with some time lag. My initial research confirmed that industry sectors were particularly likely to collocate with other "related" industries in English cities – that is, those industries with which they regularly share skills, labour and products – according to my two relatedness models [18]. The relationships were found to be highly skewed, with a few industries showing strong tendencies to interrelate and collocate and a long tail of more weakly interdependent economic sectors.

The next step was to see how this sort of interrelatedness materialised in a particular place: the city of Greater Manchester. This city provided a very interesting case study to explore. Situated in the northwest of England, it is governed by a combined authority made up of a number of local authorities: Bolton, Bury, Manchester, Oldham, Rochdale, Salford, Stockport, Tameside, Trafford and Wigan – all former town centres ringing the city, which have now been incorporated into the expanded urban fabric of the city. While administratively the conurbation is known as Greater Manchester, local people more often refer to it as Manchester, or the Manchester city region, as will also often be done here. The city is recognised internationally as having played a key role in the industrial revolution from the late 18th century, being the site of some of the first factories, and rapidly becoming a global centre for textile production. It now has a more diverse urban economy, hosting an important concentration of knowledge-based services, wholesale and retail, relative to other UK cities. Nevertheless, the city continues to harbour important industrial strands of capability and knowledge, in particular in the production of "materials" (or fabrics), which are also an important focus for the city's research and development institutions.

Supply chain relationships

Figure 3.1 shows a network model, produced in Gephi, of the likely supply chain relationships happening between industries in Greater Manchester, taken from the thesis I published in 2021. It highlights in a darker grey and bold type those industries that were particularly concentrated in this city at the time of my research (in terms of the number of local businesses in these sectors). Nodes that are closer together are more likely to share products, while the weight of the edges indicates the strength of the relationship (see Box 1).

Parts and wholes: the configuration of urban economies 23

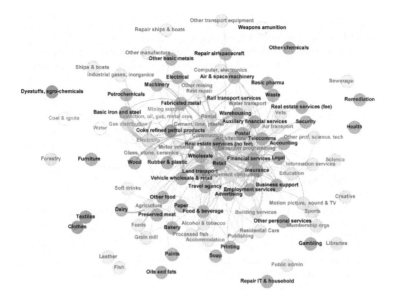

Figure 3.1 Supply chain matrix for Greater Manchester. Edges represent value of products interchanged according to UK input-output analytical tables (part of UK National Accounts) in 2014. Intensity of node shade, and shade of font, reflects relative concentration in the city according to a location quotient (based on the number of firms in Greater Manchester according to UK Business Count, 2019)

Source: Author's own diagram

Box 1 Gephi – a tool for visualising networks

Network software programs like Gephi visualise network relationships. They combine lists of nodes and lists of edges which indicate the strength of the relationship between these nodes and use these to form matrix diagrams. They therefore 'turn structural proximities into visual proximities' as a tool for analysis [26, p. 2]. Gephi uses algorithms like Force Atlas which is based on a process of attraction and repulsion. Nodes act like charged particles, which repulse each other, while the edges simultaneously attract the nodes to each other, based in part on their "edge-weight". The closeness of the nodes in the resulting matrices therefore reflects the relative strength of their ties. Different algorithms (such as GenLouvain) can then be used to detect communities in the network through identifying groups of nodes that are more closely connected to each other than to the rest of the network.

24 *Rebuilding Urban Complexity*

It is immediately apparent that this model of the potential for supply chain relationships in Greater Manchester has "hierarchy" – with groupings of industries that are likely to have particularly strong interconnections. The matrix is visually split into three main types of economic sector – services (to the bottom right) and manufacturing (to the left), while construction and logistics are in the centre. Within the services sector, supply chain relationships are relatively weak but also dense, reflecting the fact that many services are in fact "business services" with the capacity to serve a wide range of different local sectors. Financial services, wholesale and telecommunication also have a high "betweenness centrality" in the network – they are more likely to be *connectors* between one part of the network and another. The other key sector with high "betweenness centrality" is the construction sector, being at the heart of the economy when it comes to supply chain relationships. In fact, this sector has more relationships than any other with different local industrial sectors. While the construction sector is not particularly concentrated in Greater Manchester (compared with other English cities), it has strong links into areas where the city *is* concentrated, including the manufacturing of rubber and plastic, wood and retail.

Labour sharing

Figure 3.2 takes a different approach, this time focusing on skills-relatedness between industries instead of potential supply chain relationships. Again, the matrix combines local data (this time of local employment by sector) with a model predicting interrelationships (this time based on Neffke's German skills-relatedness model). In this matrix there is also a split between service sectors and the rest of the economic sectors in the city. You can see that public services, knowledge-based services and the arts and media are all relatively interconnected to the bottom left of the figure, suggesting that these sectors are likely to share a common pool of labour. In the top right centre of the network diagram, there is a clear and dense cluster around chemicals, gases, soaps, paints and coke, all of which are concentrated in Greater Manchester compared to elsewhere in the country – even if the numbers of jobs in these industries are not very high [18]. This represents a collective industrial intelligence which is nested in the city.

Looking at the economy at a finer grain

Coming back to the system concepts reviewed in Chapter 2, it is interesting to note that economic diversity in the city looks very different depending on the "scale of resolution" with which it is viewed. In these matrices, sectors are named according to the NACE industrial classification which is commonly used in Europe. This classification has a number of different levels to it – you can zoom out to look at broad industry sectors (e.g., manufacturing, services) or zoom in to industries at a very fine grain of definition (e.g., performing arts or textiles wholesalers).

When the potential interrelationships in the economy of Greater Manchester are examined at a finer grain, this produces network maps which are difficult to

Parts and wholes: the configuration of urban economies 25

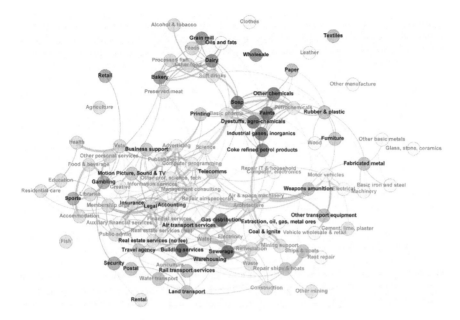

Figure 3.2 Skills-relatedness matrix for Greater Manchester. Edges represent skills overlap (according to Frank Neffke's skills-relatedness model, sourced from http://doku.iab.de/fdz/reporte/2017/MR_04-17_EN_data.zip). Intensity of node shade, and shade of font, reflects relative concentration in the city according to a location quotient (based on the number of employees in Greater Manchester according to Business Register and Employment Survey, 2018)

Source: Author's own diagram

disentangle with the naked eye. The complexity starts to become overwhelming when the interdependences that are likely to exist between more specialist subsectors are taken into account. Nevertheless, we can decipher these networks by again exploring the hierarchy inherent in the system (see Figure 3.3). Both the skills-relatedness and supply chain networks were found to split into emergent "economic communities" which formed an additional layer between the individual parts (sectors) and the urban economy as a whole. Industries within these communities were more likely to have relationships with each other, as compared with the network as a whole. We could call these subgroupings "communities of production" (industries more likely to share products in supply chains) and "skills basins" (a concept developed by Neave O'Clery and colleagues [27]). Network analysts define the ease with which one may break down networks into communities as the "modularity" of the network. In this case, the skills-relatedness matrix was found to be more modular than the supply chains matrix, implying that skills basins are more tightly defined than communities of production.

26 *Rebuilding Urban Complexity*

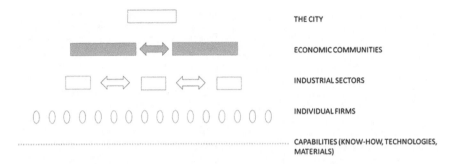

Figure 3.3 The position of economic communities in the economic hierarchy of cities
Source: Author's own diagram

Skills basins

The skills basins identified through my research in Greater Manchester are visualised in Figure 3.4. We can zoom in to explore these economic communities in more detail – to identify how the relationships within them might be topologically structured at a fine grain (more specifically, a 4-digit NACE industry classification). As previously noted, one important skills basin in Greater Manchester is that associated with knowledge-based services, with this community accounting for nearly a quarter of Greater Manchester's total workforce at the time of this study [18]. Interestingly, within this skills basin the financial and legal sectors appear to be particularly likely to share common labour pools.

When the chemicals, energy and gases skills-basin (also previously identified) is analysed at a finer grain, chemicals, dyes, paints, household sanitary goods and soap form a particularly dense cluster. According to UK Business Count data, Greater Manchester hosted 10 companies specialising in dyes and pigments at the time of my study, which was 13.3 times the Great Britain average. Also, the engineering and hard manufacturing skills basin includes people working in paper, plastic packaging and rubber manufacturing in the city, but the strongest links are between the glass and metal-related industries.

It is important to reiterate that the economic communities that I have been describing emerge from the network analysis itself. Firstly, this analysis reveals that certain industry sectors are more likely to have dense interrelationships with each other than with all the others, and then these clusters are speculatively named on the basis of their constituent industries. This is a very "bottom-up" way of understanding the interlinkages which are happening in a given urban system, as compared with just analysing the standard groupings set out in formal industrial classifications. Indeed, sometimes these emergent communities contain bedfellows which would normally be separated at a much higher level in industry classification trees, with the textiles and clothing skills-relatedness community including sectors associated with wholesale, manufacturing and retail (see Figure 3.5).

This community also reveals strong links between more traditional industries, such as textile weaving, and cutting-edge strands of advanced manufacturing, such

Parts and wholes: the configuration of urban economies 27

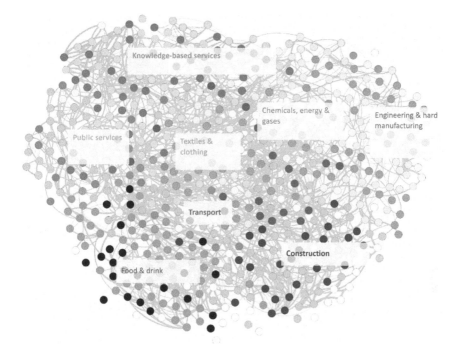

Figure 3.4 Skills basins in Greater Manchester
Source: Author's own diagram

as technical textiles. Indeed, I was told anecdotally about a local suit maker who had been experimenting with technically advanced forms of 3D weaving (involving aluminium fibres) in one corner of his suit factory. The sectors in this community offering the highest employment in Greater Manchester are clothing retail, on-line shopping and wholesale. This may not come as a surprise, given local firms like Boohoo.com and Pretty Little Thing which have come to dominate the online fast-fashion market in the UK.

Communities of production (or supply chains)

Communities of production would seem to be rather different from skills basins, considering that they offer the potential for the sharing of products, goods, materials and technologies rather than ideas and skills. We earlier discussed the skills basin associated with textiles and clothing, but these sectors also share a common "community of production" which is rather separate from the rest of the network, suggesting that firms in these sectors mainly exchange products between themselves. Even more fine-grained information about these supply chains was gathered by the Alliance Project, which carried out research with 500 actors in the textiles and clothing sector in the North West of England in 2014 [28]. These researchers

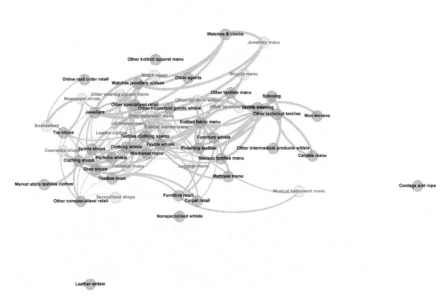

Figure 3.5 The textiles and clothing skills basin

Note: the abbreviation "whlsle" in this diagram refers to wholesale
Source: Author's own diagram

found that there was a core network of five particular industries that were still strongly interlinked in terms of their supply chains: spinners, weavers, homeware, underwear, "dyers and finishers" and retail. They also found that the textiles and clothing sectors had greater links to chemicals firms than to the digital and creative industries.

Comparing economic communities

While these fine-grained explorations of economic interlinkages can prove fascinating, we can also compare the characteristics of economic communities in a more abstract way – zooming out to see the overall likely structure of urban economic complexity in a city. For example, some of the communities in the two networks analysed here were found to be more "outward facing" while others were "inward facing" – in that they had a greater proportion of internal linkages as compared with external ones. As has already been noted, the textiles and clothing community was found to be the most inward-facing community within the supply chain matrix, while the public sector (incorporating government-funded services) was relatively inward-facing within the skills-relatedness matrix.[2] In contrast, we can

Parts and wholes: the configuration of urban economies 29

also see that certain economic communities – such as engineering and hard manufacturing, construction, and knowledge-based services – have a greater number of outward-facing linkages into other communities, suggesting that they share capabilities and skills with a broader cross-section of firms within the local economy. These sectors may therefore serve an important role in the knitting together and reproduction of the economic capabilities of the city as a whole.

Some economic communities have particularly strong internal ties in terms of their average internal edge weight. As a reminder, edge weights in networks represent the amount of interaction between two nodes – in this case, either the amount of labour flowing between industries (and hence their skills overlap) or the value of products (in the supply chain matrix). For example, the "transport" skills basin has a particularly high average internal edge weight, meaning that there are particularly strong labour-sharing opportunities between sectors within the community. The construction-based community of production similarly has strong internal connections.

It is also interesting to explore how far individual sectors are embedded within their own economic communities. We can construct "ego networks" showing how one particular node (or industry sector) is related to all the other nodes within the network. While the ego networks of some sectors span only their own economic community (with examples being textile weaving and soap manufacture), other sectors have broad, economy-wide spans. Paper manufacturing, for example, has strong edges with sectors across five different economic communities in Greater Manchester. Another locally concentrated sector providing a "skills bridge" between two economic communities in the city is paint manufacturing, which falls between the skills-related economic communities of chemicals, energy and gases and engineering and hard manufacturing when it comes to its strongest edge weights. Such sectors represent an important crossover point in capabilities, encouraging the cross-fertilisation of ideas and working methods across very different parts of the economy, with Fortunato and Hric [29] suggesting that such bridging sectors play a role in the 'dynamics of spreading processes' across networks.

Relative local embeddedness in Greater Manchester's economy

By exploring local industry data, we can identify which economic communities are *most locally* embedded in terms of how far their constituent economic sectors are concentrated in the city in comparison to Great Britain as a whole. While we have already looked at this at the scale of individual sectors, it is even more powerful to understand how *interlinked economic communities* are concentrated in particular cities. The skills basin most locally embedded in Greater Manchester was the textiles and clothing community – not surprising, given the long history which this city has had in these sectors, as we will explore in Chapter 6. The local embedding of this community may explain why the textiles sector is particularly productive compared to the rest of the country and why the city continues to innovate in the area of clothing. The chemicals, energy, and gases community was also found to be

30   *Rebuilding Urban Complexity*

relatively strongly locally embedded, with the knowledge-based services community coming third [18]. The local embeddedness of an economic community may provide a degree of resilience, in that sharing a common cross-sector labour pool means that if one particular sector declines, workers can relatively easily transition to other sectors in the community, transferring their existing skills and capabilities. However, it may also create a degree of "lock-in" and a domino effect, whereby decline in one sector has knock-on effects for its related sector – something that we will also discuss further in Chapter 6.

Common problem definition

It is worth taking a step back and thinking about the reason behind these shared capabilities which underpin Greater Manchester's economy – an under-researched topic in the economic-relatedness literature. What are the common skills and technologies which mean that people can transition easily, for example, between one sector and another? Hillier [30] suggested that knowledge spillovers are more likely between trades that share 'some common problem definition'. In Greater Manchester, one common capability which seems to be shared across several different sectors is the manipulation of *materials* – not only textiles, but also paper, metals, plastics, chemicals, rubber and plastic and comestibles such as food and drink. Within this broad area of specialism, subgroups may form because different types of materials require different types of handling. Here, the food and drink community appears to operate as a relatively separate skills basin, perhaps because this sector deals with more highly perishable materials which must be produced in very hygienic conditions. The rubber and plastic sector is particularly well embedded within Greater Manchester's economy, sharing skills and labour (and hence potentially a common "problem definition") with other industries such as textiles, paper, machinery and chemicals. In fact, this sector employed more people than did the textiles sector in 2018. Depending on your viewpoint, this might seem problematic (as these materials have had a negative environmental impact) or promising (given that this indicates a collective urban intelligence which could be at the heart of attempts to make these materials less environmentally impactful, something that we will discuss further towards the end of this book).

How firms use capabilities interchangeably

As previously identified, a central idea to relatedness research is that supply chains, labour flows and technology exchange are not such different things; they are simply ways of bringing the embedded capabilities of a city together. This idea was reinforced in interviews that I carried out with manufacturing companies in Greater Manchester and in London. As an example, the Salford-based manufacturer Wright Bower (see Box 2) draw on people, machines and suppliers somewhat interchangeably in their production chain for making leather bags. This includes heavy-weight sewing machines, specialist machines for tubular work, machines for buffing and grinding edges, and machines for adding logos to finished products. However,

Parts and wholes: the configuration of urban economies 31

pattern cutting is done by people, and staff also carry out stitching tasks as piece-work, sometimes in their own homes. The brass tools which print the logos onto the leather are created by a supplier in Southend which often does this process overnight. The decision to use one or another capability is often made based on cost. Wright Bower still do pattern cutting by hand, for example, because invest-ing in a digital cutting machine would require a significant amount of additional investment.

People also have their own advantages (as compared with machines) when it comes to flexibility. A waterproof-clothing firm in Salford – Private White V.C. – told me that they still do processes such as pattern cutting by hand, for example, because it means that they can flexibly start new product runs, easily changing the products that they make. I heard a similar attitude regarding the inter-changeability of capabilities when doing research in the Old Kent Road in London. Here a manufacturer of trays (Kaymet) showed me the array of machines which constitute the factory's production line, with the trays going from one machine to the next as they are prepared. Nevertheless, at some point in the line the trays go out of the factory altogether, for that stage of the production process to be dealt with by a supplier, before the trays then come back to continue their progression along the chain of machines.

Finally, such machines can be an important, and neglected, store of knowl-edge in cities. The philosopher Bruno Latour discusses how machines embody the knowledge of people through 'delegated programmes of action' [31]: when more programmes are delegated, machine management becomes one of the key aspects of production. This also means that in some cases machine technicians can make valuable company managers. A knitted-hat manufacturer in Salford is managed by two former machine technicians. For this firm, machines are so important to their business that they keep them as secret as possible – they would prefer that their rivals did not deduce their production processes from seeing them (see Box 4).

Comparing local embeddedness with other UK cities

How does the particular pattern of locally embedded related industries in Greater Manchester compare with the "footprints" of other UK cities? I learnt more about this through working with colleagues at the UCL Bartlett Centre for Advanced Spa-tial Analysis and the Alan Turing Institute to compare skills-relatedness in Man-chester with other UK cities, this time using a skills-relatedness model based on a 1 per cent sample of UK labour market data.[3] The neighbouring city of Sheffield, for example, has a contemporary concentration in metalworking – which is not sur-prising given its long history in cutlery making and steel – in addition to other skills basins such as finance and the production of medical instruments. Overall, this city was found to have fewer individual industry specialisms across fewer skills basins than Greater Manchester, while specialisms were more "isolated" in terms of being less likely to be surrounded by related nodes than in Greater Manchester or London. This might make Sheffield more vulnerable to economic shocks, given that people are not able to transition so easily into related industries. The West

32   *Rebuilding Urban Complexity*

Midlands (where Birmingham lies) has specialisms in industries relating to the car sector, but also has health tech. London has a particularly high concentration in service industries associated with finance, broadcasting, education, entertainment and retail. Eighty per cent of the industries in the media-related skills basin were found in the capital city.

London, Oxford and Cambridge had the most diverse sets of related industries, making these cities particularly resilient. Indeed, when this same UK-specific skills-relatedness dataset was used by colleagues to analyse the impact of relatedness on recovery after the 2008 "great recession", cities with more related industries were found to have bounced back quicker [21]. A similar finding came from an OECD study at the level of counties in the United States, where interindustry flows were found to be key to post-recession employment growth [22]. Again, this is strong evidence that resilient urban economies function as interconnected wholes as opposed to more self-sufficient individual parts.

Open cities and defining systems boundaries

The previous analysis focused on interrelationships between industries *within* cities. As discussed in Chapter 2, we also need to remember, however, that cities are very much open systems. Focusing just on the city itself would be an error of "boundary definition"– particularly as today's supply chains and labour flows have a strong global dimension. When I talked to local firms involved in textiles, clothing and accessories, for example, it was clear that they were very much involved in international supply chains and indeed production networks. Wright Bower were involved in an integrated production process which spanned the UK, Italy and Belgium (see Box 2). Such international material flows were common for all the companies I interviewed, in Manchester and in London, where it was pointed out to me that this reliance on international supply chains was partly the result of the collapse in recent years of UK manufacturing [18, 32].

Box 2   Wright Bower: participating in global production chains under the arches

Wright Bower manufacture leather bags and wallets. First established in 1976, they have since 2008 been based in railway arches [18]. Wright Bower are embedded within a very international production chain. Certain goods are sourced locally, including metal tools and synthetic fabrics. Leather comes from the UK, Italy, Belgium and Spain. Threads come from Germany, and inks and studs from Italy. The production chain goes back and forth across different European countries until the bags are ready for sale.

At the more local scale, the railway arches under which the company is based have allowed them to gradually expand. Wright Bower first rented three arches, and then knocked an entrance through into a fourth. They have also added a mezzanine floor. While the arches have come with certain problems (such as damp), they have provided an affordable set of spaces close to the city centre. Railway arches have been found elsewhere to be well set up to support processes of experimentation and modular expansion [33].

Source: The information in this Box is based on an interview carried out during my PhD research – see [18].

Labour pools are also increasingly international in origin. While all of the local companies in interviewed in Manchester were dependent on a local (sometimes very local) labour pool, many of these workers were born elsewhere. The waterproof-clothing manufacturer Private White V.C. (see Box 7) has been called the 'united nations of tailoring', having staff from Poland, Mongolia, Romania, Pakistan, Afghanistan, Spain, Syria, Kurdistan, Bulgaria, Latvia, Slovakia, the Czech Republic and Russia [34]. These workers brought important skills and also a tolerance of relatively routine job tasks.

In summary

This chapter has focused on the fact that urban economies are highly *relational*. Local workers and firms in cities are part of something bigger than themselves – a part-whole relationship based on multiple interdependences – even if cities are always embedded within wider economic networks. This part-whole structure has a hierarchy, with smaller communities likely to form economic sectors that strongly share capabilities, called here "communities of production" and "skills basins". Analysing these interrelationships provides an understanding of the embedded collective intelligence existing in post-industrial cities such as Manchester which can offer an important resource for the future.

In the next chapter we explore the configurational relationships which structure networks of urban *public space* as a basis for later exploring how economic and spatial part-whole systems work together in cities.

Notes

1 It is worth noting that skills-relatedness models have been developed for the UK through other means, such as focusing on similarity of tasks within occupational structures. See, for example, the work of Guilia Faggio and colleagues. In recent years I have worked with Neave O'Clery and colleagues at University College London and the Alan Turing Institute on a new longitudinal dataset on labour flows, based on the Annual Survey of Hours and Earnings. However, as this was based on a 1 per cent sample it did not possess the detailed modelling capability of Neffke's German dataset, so I use it only for comparative purposes here. Comparisons between national skills-relatedness models have revealed that there are similarities in the patterns of skills-relatedness between industries in the UK and in other European countries – although both the UK and Ireland were found to have a more fragmented and modular structure, nested within the larger clusters

34 *Rebuilding Urban Complexity*

of networks also found in Germany, Sweden and the Netherlands – suggesting that there is less labour mobility across skills basins in the UK and Ireland. See Straulino, D., M. Landman, and N. O'Clery, A bi-directional approach to comparing the modular structure of networks. *EPJ Data Science*, 2021. **10**(1): p. 13.

2 An alternative way of identifying the relative integration of different economic communities into the whole network would be the "random walk method", which successively merges communities according to the strength of their interconnections.

3 The dataset is based on a sample from the Annual Survey of Hours and Earnings (ASHE) which provides employment information for 1 per cent of jobs in the HM Revenue & Customs Pay As You Earn (PAYE) records.

References

1. Chandler, D., Beyond neoliberalism: resilience, the new art of governing complexity. *Resilience*, 2014. **2**: p. 47–63.
2. Marshall, A., *Principles of economics*. 1890, London: Macmillan.
3. Marshall, A., *Industry and trade*. 1919, London: Macmillan.
4. Duranton, G. and D. Puga, Micro-foundations of urban agglomeration economies. In *Handbook of regional and urban economics*. 2004, Amsterdam: Elsevier. P. 2063–2117.
5. Harvey, D., Reshaping economic geography: the world development report 2009. *Development and Change*, 2009. **40**(6): p. 1269.
6. Krugman, P., What's new about the new economic geography? *Oxford Review of Economic Policy*, 1998. **14**(2): p. 7–17.
7. Martin, R. and P. Sunley, Complexity thinking and evolutionary economic geography. *Journal of Economic Geography*, 2007. **7**(5): p. 573–601.
8. Sunley, P., Relational economic geography: a partial understanding or a new paradigm? *Economic Geography*, 2008. **84**(1): p. 1–26.
9. Juhasz, S., Z. Elekes, and J. Gyurkovics, Network revolution in economic geography. In *New ideas in a changing world of business management and marketing*. 2015, Szeged: University of Szeged.
10. Neffke, F., A. Otto, and A. Weyh, *Inter-industry labor flows*. IAB-Discussion Paper 21/2016. 2016, Nürnberg: IAB, Federal Employment Agency.
11. Boschma, R. and S. Iammarino, *Related variety and regional growth in Italy*. SPRU Electronic Working Paper Series. 2007, Brighton: University of Sussex.
12. Frenken, K., F. Van Oort, and T. Verburg, Related variety, unrelated variety and regional economic growth. *Regional Studies*, 2007. **41**(5): p. 685–697.
13. Hidalgo, C.A., et al., The principle of relatedness. In *Proceedings of the international conference on complex systems*. 2018, Cham: Springer.
14. Neffke, F., M. Henning, and R. Boschma, How do regions diversify over time? Industry relatedness and the development of new growth paths in regions. *Papers in Evolutionary Economic Geography*, 2009. **9**(16).
15. Ellison, G., E.L. Glaeser, and W.R. Kerr, What causes industry agglomeration? Evidence from coagglomeration patterns. *American Economic Review*, 2010. **100**: p. 1195–1213.
16. Diodato, D., F. Neffke, and N. O'Clery, Agglomeration economies: the heterogeneous contribution of human capital and value chains. In *Papers in evolutionary economic geography 16.26*. 2016, Utrecht: Urban & Regional Research Centre, Utrecht University.
17. Faggio, G., O. Silva, and W.C. Strange, Tales of the city: what do agglomeration cases tell us about agglomeration in general? *Journal of Economic Geography*, 2020. **20**(5): p. 1117–1143.
18. Froy, F., *'A marvellous order': how spatial and economic configurations interact to produce agglomeration economies in Greater Manchester*. Bartlett School of Architecture. 2021, London: University of London.

Parts and wholes: the configuration of urban economies 35

19. O' Clery, N., A. Gomez-Lievano, and E. Lora, *The path to labour formality: urban agglomeration and the emergence of complex industries*. Working Paper No. 78. 2016, Cambridge: Centre for International Development at Harvard University.
20. Tsvetkova, A., T. Conroy, and J.-C. Thill, Surviving in a high-tech manufacturing industry: the role of innovative environment and proximity to metropolitan industrial portfolio. *International Entrepreneurship and Management Journal*, 2019: p. 1–27.
21. Straulino, D., D. Diodato, and N. O'Clery, *Economic crisis accelerates urban structural change via inter-sectoral labour mobility*. 2022, Seville: Joint Research Centre (Seville site).
22. Partridge, M. and A. Tsvetkova, *Local ability to rewire and socioeconomic performance: evidence from US counties before and after the great recession*. OECD Local Economic and Employment Development (LEED) Papers 2020/04. Paris: OECD.
23. Neffke, F., *Regional skill-bases: identifying the local labour force's diversification potential*. 2010, Brussels: DG Regio Open Days.
24. Otto, A. and A. Weyh, *Industry space and skill-relatedness of economic activities: comparative case studies of three Eastern German automotive regions*. 2014, Nürnberg: IAB-Forschungsbericht.
25. Hidalgo, C.A., *Why information grows: the evolution of order, from atoms to economies*. 2015, London: Penguin Books.
26. Jacomy, M., et al., ForceAtlas2, a continuous graph layout algorithm for handy network visualization designed for the Gephi software. *PLOS One*, 2014. **9**(6): e98679.
27. O'Clery, N. and S. Kinsella, Modular structure in labour networks reveals skill basins. *Research Policy*, 2022. **51**(5): 104486.
28. The Alliance Project, *Supply chain mapping and targeting the investment pipeline*. 2016, Manchester: The Alliance Project.
29. Fortunato, S. and D. Hric, Community detection in networks: a user guide. *Physics Reports*, 2016. **659**: p. 1–44.
30. Hillier, B., The fourth sustainability, creativity: statistical associations and credible mechanisms. In *Complexity, cognition, urban planning and design*, J. Portugali and E. Stolk, Editors. 2016, Cham: Springer International Publishing.
31. Latour, B., Where are the missing masses? The sociology of a few mundane artefacts. In *Shaping technology-building society: studies in sociotechnical change*, B. Wiebe and J. Law, Editors. 1992, Cambridge, MA: MIT Press, p. 225–259.
32. Domenech, T., F. Froy, and N. Palominos Ortega, *The maker-mile in East London: case study report*. 2019, Brussels: Cities of Making.
33. Froy, F. and H. Davis, Pragmatic urbanism: London's railway arches and small-scale enterprise. *European Planning Studies*, 2017. **25**(11): p. 2076–2096.
34. Lancashire Life, *Private White VC – the designer label inspired by a Salford war hero*. 2014. https://www.greatbritishlife.co.uk/lifestyle/fashion/22620755.private-white-vc-designer-label-inspired-salford-war-hero/. Accessed [10/09/2024].

4 Parts and wholes

The configuration of urban space

In order to understand how urban space works as a system, we have to rethink our commonly held understandings of space and consider the topological relationships which link together streets and urban public spaces. In topological considerations of space, the quality of a part (e.g., a square or street) can be understood only through its configurational relationship to the network as a whole. The geographer Helen Couclelis [1] describes how such a topological understanding of space differs from Euclidean models which measure space according to metric distance. Topological relationships remain the same despite distortions in metric distance, like a net or a piece of knitting which retains the same structure despite being stretched.

Such understandings of space have gained greater traction in recent years as people have been able to plot and analyse networked relationships using computer software. However, architectural interest in the interconnections which exist between public spaces goes much further back. In 1889, Camillo Sitte became fascinated, for example, by the layout of medieval trading cities, exploring how public spaces in the Italian cities of Padua, Syracuse and Palermo privileged interconnections and movement across the city [2]. Churches were set back from the main square or public space, so that they did not obstruct the flow of people from one part of the city to another. Public spaces were also often triangular in shape, created as a "by-product" of a broader linear pattern of streets which encouraged the pedestrian movement that supported encounter, trade and exchange.

In more recent times, the importance of linear spatial pathways through cities was also recognised by the American urban thinker Kevin Lynch, who published the oft-cited *The Image of the City* in the 1960s. Lynch investigated how people gain a sense of the whole city in which they live, as opposed to just their own local part of it. He identified a set of urban forms which help cities to become intelligible to the people who live in them, including "nodes", "edges", "districts" and "landmarks". He emphasised, however, that 'the paths, the network of habitual or potential lines of movement through the urban complex, are the most potent means by which the whole can be ordered' [3, p. 96].

Lynch's contemporary Jane Jacobs also understood how important street networks are in connecting parts of cities into functioning wholes. She pointed out that it was impossible to understand a part of the system (for example a park)

DOI: 10.4324/9781003349990-6

Parts and wholes: urban space   37

without understanding how it fitted into broader patterns of urban interconnection and movement, and suggested that this was crucial in ensuring that public spaces in cities are well used. As she said,

> How much the park is used depends, in part, upon the park's own design. But even this partial influence of the park's design upon the park's use depends, in turn, on who is around to use the park, and when, and this in turn depends on uses of the city outside the park itself. [. . .] Increase the park's size considerably, or else change its design in such a way that it severs and disperses users from the streets about it, instead of uniting and mixing them, and all bets are off.
>
> [4, p. 565]

Elsewhere she talked about an interconnected underlying spatial "anatomy" which shapes city life [see 5].

Space syntax: a tool for understanding spatial complexity

Each of these thinkers had an influence on the late Bill Hillier, who worked with colleagues at the Bartlett, University College London, for more than 50 years to put configurational part-whole ideas of space on a more scientific footing. Hillier pointed out that where there is scale there is also generally structure – and in the case of cities there is a fine-grained configurational structure which allows cities to host large densities of people, while preserving intelligibility and accessibility throughout the urban system. While many people think of architects as being principally focused on the design of buildings, Hillier was primarily interested in the networked properties of the space *in between* the buildings.[1] He and his erstwhile colleague Julienne Hanson argued that it is 'this ordering of space that is the purpose of building, not the physical object itself' [6, p. 1]. Elsewhere he suggested that cities were 'things made of space' [7, p. 262]. Hillier and Hanson pointed out that a key property of a street is that 'it is a unique and distinguishable entity, yet at the same time is only such by virtue of its membership of a much larger system of spatial relations' [6, p. 79]. In particular, the amount of pedestrian movement in a particular street is determined by its accessibility to all the other streets in the system. Hillier found that while urban street networks can only shape movement patterns probabilistically,[2] the network patterning of the space between buildings was the most powerful single determinant of urban movement, both pedestrian and vehicular [7]. He termed the movement which is shaped by space itself 'natural movement'.

Describing the network properties of urban public space is not easy, as this space is often continuous and not broken up into discrete shapes. It is particularly hard to describe how urban spaces relate not only to the immediate spaces around them but also to wider spatial networks [8]. To understand these complex relationships, space syntax analysts draw on graph theory (a network-based theory used in mathematics) and use computer programmes such as "UCL Depthmap".[3] With

38   Rebuilding Urban Complexity

this type of software, the now international space syntax community has gathered a significant amount of evidence over the past fifty years about the correlations between the networked characteristics of urban space and vehicular and pedestrian movement patterns in cities [see, e.g., 9].

While there are similarities between space syntax and other types of graph theory,[4] space syntax analysis incorporates a particularly "architectural" understanding of networks. It has been described as allowing architects to "access complexity", in a way which is comprehensible to intuition [10]. Bill Hillier and colleagues were preoccupied with the lived experience of people as they move through the urban system, seeing our experience of cities as being strongly shaped by practices of 'seeing and going' [11]. Streets provide "lines of sight" which offer intelligibility to people as they navigate through space. The subtle ways in which cities structure space to support navigation are not necessarily understandable when looking at the structure of a city "top down", using a map. Indeed, Hanson [12] makes a distinction between the *structure* of the urban environment (where what is important is intelligibility for the embodied navigator) and the *order* depicted in maps and spatial plans (which is about intelligibility as a conceptual scheme to be understood by the external observer). In his popular book *The Practice of Everyday Life*, Michel de Certeau likewise compares our engaged way of navigating urban streets with the disembodied view that people have as they look down on the city from above. He writes that when someone looks down on New York streets from a skyscraper, for example, their 'body is no longer clasped by the streets that turn and return it according to an anonymous law', and goes on to identify that the person's 'elevation transfigures him into a voyeur. It puts him at a distance. It transforms the bewitching world by which one was "possessed" into a text that lies before one's eyes' [13, p. 92].

As we will see later in this book, when parts of cities are explicitly "designed" and master-planned, planners and architects often impose a "geometric" order on cities – involving, for example, circles and semicircles – without adequate consideration for what it will be like to physically move through such spaces on the ground. Thinking about cities in terms of the experience of moving through them also reveals the importance of different vantage points, or different positions within the urban fabric – how you experience the city depends very much on your own position within it. This is something we will also return to.

Two key space syntax variables

Space syntax analysis focuses on two specific aspects of the configuration of cities. The first is the "depth" in space of each street from all the others in the system – how many turnings does it take to get from each street to all the others? When a street is relatively "shallow" or easily accessible to those around it, it is identified as being "spatially integrated" – with the overall variable used in this analysis in Depthmap called "integration". The concept in broader network analysis most similar to this is "closeness centrality", and indeed this form of analysis is often used to identify particularly integrated parts of cities that form local or city centres.

The other main type of space syntax analysis focuses on the parts of the city that are most likely to host "through movement". The associated Depthmap measure of "choice" calculates how far a street (or street segment) falls on the shortest path linking any other pair of segments in the street system. Again, the "shortest paths" are defined by the number of turnings and degree of angular change from origin to destination. This measure is similar to the broader network concept of "betweenness centrality" in that it looks at those network elements that are most likely to gain importance through being "in between" other network elements.

The dual system of city streets

Bill Hillier described how cities have dual systems – a foreground network of higher "choice", better-connected streets and a background network of streets which are less connected. The foreground network of streets generally comprises a small percentage of the wider network, but it is key in channelling movement around the city. It is made up of relatively long streets which require people to make few angular changes as they traverse the system, and which hence provide Kevin Lynch's 'lines of movement through the urban complex' – increasing the overall intelligibility of the city. The foreground network of streets also increases the speed with which people can go from one part of the city to the next – crucial, for example, for the daily commute. These streets often reach from the city centre towards the outskirts of the city in what Hillier called a 'deformed wheel' or hub-and-spoke structure. Conversely, background streets are often shorter and at right angles to each other, and constitute the quieter residential areas of the city. While the foreground network of cities plays a role in connecting and "gathering together" streets into a functioning whole, the "background network" of shorter streets also plays a role in knitting together the city. Hanson [12, p. 297] describes the 'process of fine-tuning of the urban grid by which the whole comes to dominate the parts'. Where cities evolve organically or incrementally over time, generally backstreets are only one or two turnings away from a busier foreground network street, giving people access simultaneously to both their local place and the wider city, something Bill Hillier calls a 'two-step logic'.

Pervasive centrality

Space syntax theory has also built on Jane Jacobs' perception that street networks operate as *multi-scale* systems in cities. While urban transport analysts often become preoccupied with understanding distances and times from origins to destinations, space syntax considers the *by-products* of such movement, with patterns of through movement in cities at different scales creating an important 'layering of movement networks' [14]. Whereas some people may be using a street to go to a local bakery or take their children to school, other people will be using the same street as part of a longer trip to work, or passing through on their way to meet a friend in another part of the city. The combination of these movement paths contributes to the overall co-presence of people created by the street. In order to reveal

40 Rebuilding Urban Complexity

the particular patterns of accessibility which emerge through different scales of movement, space syntax analysis explores connectivity at a number of different radii – from the local areas which would be accessible by foot (within a 400–800 metre radius) up to larger areas of the city that might be traversed using vehicles (e.g., 2,000 metres) and then for the whole of the city (which in space syntax analysis is given the classification "Rn").

Commercial activities are often supported in places of what Jane Jacobs called "confluence and convergence" in street systems, with small urban sub-centres being an emergent effect of the crossings of differentially scaled networks [4, 15]. Hillier proposed that such crossings are in fact present at multiple scales in cities, creating a situation of "pervasive centrality". This is a rather different concept to that of "polycentricity", which is often used by economists to refer to the multiple sub-centres which cities host. The concept of pervasive centrality reveals that these centres are *interlinked* through – and indeed formed by – intersecting networks [16, 17]. Hillier identified that local centres often combine local network intersections and a degree of local grid intensification – the network of streets is denser in these places, decreasing the metric distance that people need to walk to access these centres. However, such centres are rarely separated from the rest of the network by clear boundaries, with a key characteristic of successful cities being that they are able to combine area differentiation with *spatial continuity* [18]. While cities may support local "niches" – to use an ecological term – this occurs without a closing off from broader systems circuits. Local centres are always linked into a larger whole. Indeed, Hillier argued that architects and urban planners often get fixated on designing local "places" without realising that 'places are not local things. They are moments in large-scale things, the large-scale things we call cities. Places do not make cities. It is cities that make places' [7, p. 112].

Many of the architects and planners of the mid-20th century also failed to appreciate this fact, trying mistakenly to create local communities through *cutting them off* from citywide movement flows in self-contained "estates", protected through minimal connecting streets and a liberal use of gates and fences [19]. Such ideas had considerable influence on the urban morphology of residential areas in the post-industrial cities of Manchester, Sheffield and Newcastle. They were reinforced by an American architect and city planner, Oscar Newman, who developed a theory of "defensible space", published in the 1970s, suggesting that safety was produced by eliminating strangers from residential areas (although in fact it is now known that the opposite is the case – there is safety in numbers [20–21]).

Cities as partially ordered and open spatial systems

Space syntax analysts are rarely prescriptive or normative in identifying how a city should be laid out, recognising that the basic principles that underly spatial configurations only ever create partially ordered systems, and that different cities incorporate very different "spatial cultures". However, Bill Hillier did share a belief with other systems thinkers, such as Mary Douglas, that there is a set of *constraints* on what is workable or *feasible* when it comes to liveable human systems – and in this case in what constitutes a liveable set of urban layouts. Biologists

Parts and wholes: urban space 41

similarly talk about the idea of "morphospaces" – there are only a specific number of arrangements in any situation which are viable [22]. As we will discuss in Chapter 9, arguably some of the geometrically designed planned parts of cities introduced in Britain in the 1960s fell outside this "feasible" range.

Finally, we again need to acknowledge the relative *openness* of the urban systems that we are describing. While cities host local structure, they also link to much broader spatial systems. Because the streets that form part of foreground street networks are generally connected to much larger regional and national street networks, they ensure that city streets offer not just density but also *extension*. Indeed, Hillier suggests that the "deformed wheel" structure of many organically grown cities develops due to the need to provide strangers from outside the city with short routes into the centre. If this structuring mechanism were absent, the centre would in fact be very "deep" and relatively inaccessible to the outside, making it unattractive as a space of trade and exchange. The very internal structuring of cities is therefore about producing "reach" (both for residents to "reach out" and strangers to "reach in"). This again speaks to the suggestion of Read et al. that urban structures work by 'bringing the potentials of the world to hand' [23, p. 16].

The spatial configuration of Greater Manchester

So how can we apply these ideas about the configurational properties of space and spatial part-whole structures to Greater Manchester? Can similar properties of space be found in neighbouring post-industrial cities such as Sheffield? The remainder of this chapter provides a brief snapshot of the properties of these two cities, which will then be explored further in subsequent chapters as we start looking at questions of spatial evolution.

A city with a dense core and a less dense inner ring

The Combined Authority area of Greater Manchester spans 1,276 square kilometres and it had a population of about 2.87 million residents in 2021, being the third largest city in the UK, after London and Birmingham. The city has relatively low density as compared to London, particularly in the belt of formerly industrial land which surrounds the city centre. The city also has a "green belt" (which was finalised as a continuous joined area in 1984) which defines its urban extent.

Spatial integration

What can we say about the spatial configuration of the city's streets? In Figure 4.1, the street network for Greater Manchester is analysed in terms of its local integration, that is, the relative depth of each street segment to all other segments within a two-kilometre radius from it. This image uses the conventional rainbow-based space syntax colour hierarchy, with street segments being coloured on a scale from red (highly integrated) to orange, yellow, green and then blue (the most segregated). A "core" of highly integrated streets becomes clear at the heart of the city, mainly located in the Manchester local authority, which extends outwards

42 *Rebuilding Urban Complexity*

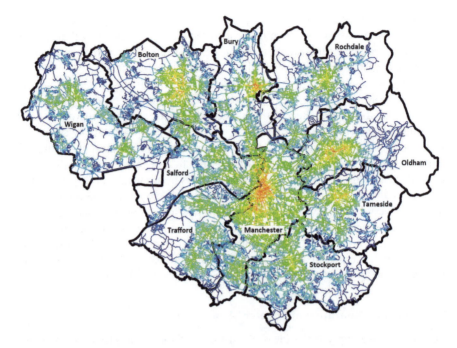

Figure 4.1 Local spatial integration in Greater Manchester

Source: Author's own diagram. Street network extracted from Space Syntax OpenMapping (see https://spacesyntax-openmapping.netlify.app/#13/51.5131/-0.1422, accessed [12/09/2024]), and amended by author. Local authority boundaries are derived from the December 2017 clipped boundaries dataset downloaded from www.data.gov.uk. Contains public sector information licensed under the Open Government Licence v3.0

towards the north and south. The Northern Quarter, to the northeast of the centre, well-known today as a buzzy centre for cafés and bars, is very well integrated at the local scale. In comparison, there are more segregated "blue" areas of urban fabric close to the urban core, and in particular in the less dense inner-city ring which may be disrupting movement patterns close to the heart of the city. Another large area of segregated "blue" at the local scale is the Trafford Industrial Park which is only weakly connected to its surrounding urban fabric (see Figure 4.2). Indeed, this industrial park links better into regional and national road systems than it does to the neighbouring city centre.

The surrounding local authorities of Bolton, Bury, Oldham and Tameside, and to a lesser extent Stockport, Wigan, Rochdale and Trafford, all show relatively integrated local centres – revealing the particularly accentuated form of "pervasive centrality" that would be expected for a city which has effectively "swallowed up" a set of surrounding towns as it expanded. Former villages which have become integrated into the broader urban fabric, among them Altrincham, Chorlton and Didsbury, also show up as being particularly locally integrated today, and are popular for wealthy residents of the city. At a still more local radius of 800 metres (about the distance people would normally walk to access local services), areas of

Parts and wholes: urban space 43

Figure 4.2 The configuration of Trafford Industrial Park compared with the city centre

Source: Author's own diagram. Street network extracted from Space Syntax OpenMapping and amended by author

"griddy" terrace streets also show up as being locally integrated in, for example, Old Trafford and Moss Side. Nevertheless, postwar planning has led other parts of the background network to become more hierarchical and "tree-like" in their internal structure, segregating areas from citywide movement flows.

The foreground network and "dual system" of streets

Figure 4.3 shows the streets in Greater Manchester that can be identified as forming the "foreground network" of the city, which clearly follows the "deformed wheel" structure which Hillier identified as being characteristic of organically grown cities. As a reminder, these are the streets that are most likely to attract through movement as people move across the urban system as a whole. This network has been extracted from the broader network of streets and superimposed on a map of building densities in the city. As can be seen in the figure, the foreground network reaches into, and connects, local centres of higher building density, with Hillier [7, p. 126] arguing that density often develops where the configuration already supports encounter. Whether the density or the spatial configuration came first, there does seem to be a correlation between density and spatial configuration in Greater Manchester, with denser parts of the urban fabric being well connected by the foreground network. At a larger scale, the foreground network also provides reach into a regional and national network of roads which connects the city to Liverpool to the west, Leeds and Sheffield to the east, and down via Birmingham to the south of the country.

The spatial configuration of Sheffield

In the neighbouring city of Sheffield, the integration core of the city also extends down to the south (see Figure 4.4) – indeed, the highest integration values are found around the Moor (a shopping street), as opposed to the area designated 'the Heart of the City' by a series of regeneration projects (something that we will pick up again in Chapter 11). This part of the city centre does indeed see high pedestrian numbers. When it comes

44 *Rebuilding Urban Complexity*

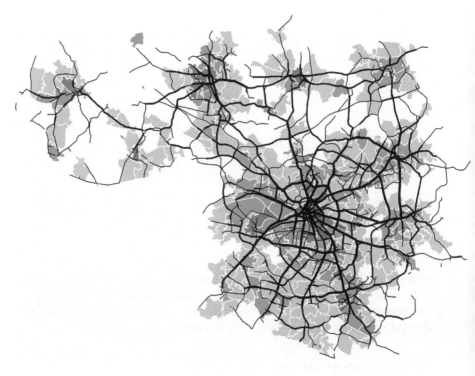

Figure 4.3 The foreground network overlain on a map of building densities. Here the foreground network constitutes the top 12.5 per cent of all streets in terms of choice or betweenness centrality. The data on building densities is aggregated by MSOA (middle layer super output area) census boundaries

Source: Author's own diagram. Street network extracted from Space Syntax OpenMapping and amended by author. Middle Layer Super Output Area Boundaries Shapefile, 2011 downloaded from www.data.gov.uk. Contains public sector information licensed under the Open Government Licence v3.0

to local integration and through movement, however, the historic city centre, with its cathedral area and Fargate, comes out strongly. While some of Sheffield's background residential network is quite "griddy", this contrasts with cul-de-sac-style developments; geometric structures (such as circles), and high-rise flats set in open space and poorly connected back into their surrounding urban fabric, including the Park Hill estate just to the east of the city centre. The foreground network of Sheffield is dominated by an inner-city ring road, with large roundabouts at junctions. It therefore has a less obvious "deformed wheel" structure. Nevertheless, there are two foreground network streets which go down through the city centre, on either side of the Moor, again reinforcing this as a topological or configurational core of the city.

In summary

This chapter has explored how urban spatial systems can be understood "configurationally", with complex multiscale networks allowing cities to work as part-whole

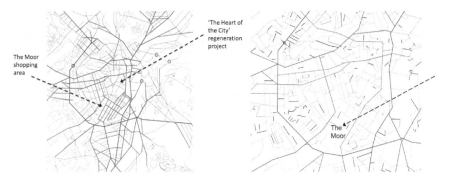

Figure 4.4 Sheffield – locating the "heart of the city"? a. Darker lines here show more globally integrated streets in the city centre. b. Darker grey streets are the ones that are expected to host most through movement

Source: Author's own diagram. Space syntax analysis based on the Space Syntax OpenMapping resource and amended with the help of Merve Okkali Alsavada, PhD researcher at UCL.

systems. The different dimensions of space syntax theory were introduced and then applied to the cities of Greater Manchester and Sheffield, which revealed not only fine-grained systemic relationships but also the imposition of other forms of order on the city. We will come back to the spatial characteristics of these and other cities in future chapters, particularly in Part 2, when we look at how these different urban morphologies evolved and changed. In the next chapter, however, we explore the links between the spatial part-whole relationships just examined, and the economic part-whole relationships analysed in Chapter 3, to identify what can be learnt by bringing together these two dimensions of complexity.

Notes

1 Bill Hillier was also interested in the space *inside* buildings, as can be seen in his many articles on the configurational structure of buildings such as the Tate Britain art museum, in London.
2 I.e., having a statistical effect once all the different journeys from specific origins to specific destinations in cities have been considered.
3 See https://github.com/SpaceGroupUCL/depthmapX/
4 Space syntax is not the only discipline to apply graph theory to the understanding patterns of urban space (see, for example, the work of Sergio Porta and Andres Sevtsuk) with different branches of graph theory disagreeing about appropriate methodology (see, e.g., Pafka et al. [2018], Limits of space syntax for urban design: axiality, scale and sinuosity). Systems thinking and complexity theory has also been applied somewhat differently to cities by Mike Batty and his team from the Centre for Advanced Spatial Analysis at University College London. This group of researchers is more likely to draw on physics rather than the embodied experience of architecture for inspiration. Batty, for example, explores the importance of fractals in the multiscale organisation of the built environment, and the importance of "scaling" within complex urban systems.

References

1. Couclelis, H., Space, time, geography. *Geographical Information Systems*, 1999. **1**: p. 29–38.

46   Rebuilding Urban Complexity

2. Sitte, C., *The birth of modern city planning [2006 edition]*. 1889, New York: Dover Publications.
3. Lynch, K., *The image of the city*. 1960, Boston, MA: MIT Press.
4. Jacobs, J., *The death and life of great American cities [1993 edition]*. 1961, New York: The Modern Library.
5. Wetmore, J.Z., *Jane Jacobs – speech at the national building museum after receiving the Vincent Scully prize*. 11 November 2000. https://academic.oup.com/cjres/article-pdf/16/3/495/52823779/rsad024.pdf. Accessed [04/10/2024].
6. Hillier, B. and J. Hanson, *The social logic of space*. 1984, Cambridge: Cambridge University Press.
7. Hillier, B., *Space is the machine: a configurational theory of architecture*. Vol. 2007 e-print. 1999, Cambridge: Cambridge University Press.
8. Hillier, B., Capturing emergence. In *Digital tsunami: VR and theory*, R. Conroy and N.S.D. Dalton, Editors. 10 February 1999. https://dezignark.com/blog/digital-tsunami-vr-and-theory-richard-barbrook-neil-dalton-bill-hillier/. Accessed [04/10/2024].
9. Lerman, Y., Y. Rofe, and I. Omer, Using space syntax to model pedestrian movement in urban transportation planning. *Geographical Analysis*, 2014. **46**: p. 392–410.
10. Davis, H. and S. Griffiths., *Bill Hillier, Christopher Alexander and the representation of urban complexity: their concepts of 'pervasive centrality' and 'field of centres' brought into dialogue*. 13th International Space Syntax Symposium. 2022, Western Norway.
11. Hillier, B., *The architectures of seeing and going: or, are cities shaped by bodies or minds? And is there a syntax of spatial cognition?* 4th International Space Syntax Symposium. 2003, London.
12. Hanson, J., *Order and structure in urban space: a morphological history of the City of London*. Bartlett School for Architecture and Planning. 1989, London: University of London.
13. de Certeau, M., *The practice of everyday life*. 1988, Berkeley: University of California Press.
14. Read, S. and L. Budiarto, *Human scales: understanding places of centring and de-centring*. 4th International Space Syntax Symposium. 2003, London.
15. Read, S., Intensive urbanisation: levels, networks and central places. *The Journal of Space Syntax*, 2013. **4**(1).
16. Hillier, B., What are cities for? And how does it relate to their spatial form? *The Journal of Space Syntax*, 2016. **6**(2): p. 199–212.
17. Hillier, B., *Spatial sustainability in cities: organic patterns and sustainable forms*. 7th International Space Syntax Symposium. 2009, London.
18. Hillier, B., The fourth sustainability, creativity: statistical associations and credible mechanisms. In *Complexity, cognition, urban planning and design*, J. Portugali and E. Stolk, Editors. 2016, Cham: Springer International Publishing.
19. Hanson, J. and B. Hillier, The architecture of community: some new proposals on the social consequences of architectural and planning decisions. *Architecture & Behaviour*, 1987. **3**(3): p. 251–273.
20. Newman, O., *Defensible space: crime prevention through urban design*. 1973, New York: Collier Books.
21. Hillier, B. and O. Sahbaz, Safety in numbers: high-resolution analysis of crime in street networks. In *The urban fabric of crime and fear*, V. Ceccato, Editor. 2012, Cham: Springer.
22. Budd, G.E., Morphospace. *Current Biology*, 2021. **31**(19): p. R1181–R1185.
23. Read, S., et al., *Constructing metropolitan landscapes of actuality and potentiality*. 6th International Space Syntax Symposium. 2007, Istanbul.

5 Bringing together configurational analysis of economies and space

How can we usefully bring together thinking about the part-whole relationships in urban space with the part-whole relationships of urban economies? What might be the role of street networks in bringing together different parts of the economy – the economic capabilities that are exchanged through labour markets, supply chains and knowledge transfer? Even though it would seem to be common sense that spatial and economic complexity reinforce each other, can we be more precise about this relationship? This would seem to be particularly important, given that spatial complexity is rarely fully acknowledged in the economic-complexity literature [1].

The Manchester hive

There has long been a recognition amongst Manchester residents of the ways that the city, as a spatial structure, brings people together to produce an economic entity that is much greater than the sum of its parts. Images of the worker bee appear on walls, in shop windows, and as sculptures, throughout the city (see Figure 5.1). The bee is celebrated as a symbol of the workers who have contributed to the economic success of the city since well before the industrial revolution – with the bee receiving new symbolic meaning as an image of social solidarity since the bomb attack at the Ariana Grande concert at Manchester Arena in 2017. Associated with the worker bee is the hive, and indeed in Victorian times, Manchester was known as the "hive of industry". While city spatial structures are clearly not the same as hives, the idea of an interconnecting spatial network of the city is clearly present in the popular imagination.

Cities have structure, not just size and density

The idea that cities might be brought together by an underlying "spatial lattice" is more rarely identified by economists. In economic theory, cities are often described according to their size and density, with less attention paid to how the spatial "parts" of a city (its streets and public spaces) fit together as a whole. Despite the fact that there has been a parallel rise of interest in complexity in the economic and

DOI: 10.4324/9781003349990-7

48 Rebuilding Urban Complexity

Figure 5.1 Bee imagery in Manchester
Source: Photographs by the author

architectural disciplines, there has been a lack of research into how urban economies might be supported by configurational *spatial structure*s [1].

This may be a reflection of the spatial concepts that economists have traditionally worked with. The word "agglomeration", for example, is a relatively crude term, which has been defined as 'the action or process of collecting in a mass'.[1] Indeed, the London School of Economics academic Henry Overman and his colleagues refer to the importance of '*economic mass*' in their analysis of the economic productivity of Greater Manchester [2]. In economic geography literature, there is also a strong focus on the "clustering" of industries at different scales, with both of these concepts (agglomeration and clustering) embodying the Newtonian idea that space might support the gravitational pull of matter into density [3]. A similar idea can be found within economic models which characterise cities as concentric rings, based on attraction to the core – such as the bid-rent theories which derive from the ideas of Von Thünen and Alonso. These models predict that economic activities will position themselves according to a downward slope of declining rental costs from the core to the periphery.

Characterising cities as merely points of gravity means that the spatial configuration of cities – so important to the generation of spatial part-whole relationships – is often overlooked. Indeed, theorists such as Ed Glaeser have explicitly argued against there being a link between the structure of the built environment and economic prosperity, with Glaeser suggesting that too much attention is paid to built structures in cities. He argues that 'cities aren't structures; cities are people' [4, p. 7], and 'cities are the absence of space between people' [4, p. 9].

Configurational analysis of economies and space   49

However, space syntax analysis has revealed a fine-grained structuring of economic activities in cities which is rooted in the network properties of street systems [see, for example, 5–10] and which influences a number of economic outcomes including levels of productivity, incomes, patenting rates and house prices [11–14]. Economic activities are also found to harness patterns of urban accessibility that are much more complex than the simple core-periphery models associated with bid rent curves [see e.g. 15]. In particular, many economic sectors locate themselves in proximity to the "deformed wheel" structure of the foreground network and its linear spokes that radiate out from the core.

Different industries seek accessibility via this foreground network in different ways. While retailers seek passing trade and "footfall" on the foreground network itself (on "the high street", the "main street" or in a town or city centre), manufacturers often seek accessibility "one step removed" in the back streets adjacent to the foreground network. In London, for example, Fiona Scott identified that industry is often found 'behind the façade, down alleys and side streets and mews', with 42 per cent of land within 200 metres of the high street in Tottenham being industrial, rising to 45 per cent in Redbridge [16, p. 208].

Bill Hillier was keen to point out that the network properties of cities only ever produce a "partial ordering" of economic activities in cities, with there being a high degree of randomness overall in how individual businesses are located. The network properties of street systems only ever serve to *limit* randomness, as opposed to constituting a strict ordering principle. Some space syntax theorists suggest that the jumbling up of different economic activities in cities contributes to a greater cohesion and functional interdependence at the global scale [see, e.g., 10, 17] – helping the parts to create a more coherent whole. Sam Griffiths has found, for example, that when it came to the early artisanal metal industries in Sheffield, it mattered less where individual activities were located in the city, as long as there was an overall coherence of industrial organisation at the urban scale created by the foreground network of streets – 'the persistence of this mechanism ensured that a high degree of randomness of location with regard to any given practitioner did not equate to a 'chaos' but rather to an information-rich structure of organized complexity' [10, p. 143].

Nevertheless, cities also offer "spatial niches" hosting more specialised industry clusters which evolve and persist through more local scale agglomeration effects. Later in this chapter we will explore the fashion cluster which occupies a particular locational niche in Strangeways in north Manchester. Another strong cluster in Manchester's history was the hat-making industry in Stockport (scc Box 3) [18]. London also has many local spatial niches which support local agglomeration economies in specialised industries, like furniture in Tottenham Court Road, tailoring (Savile Row) and South Indian restaurants (in Drummond Street near to Euston station). While local agglomeration and clustering effects are obviously important to these specialised areas, there is a *configurational underpinning* to these clusters – in both how they are spatially arranged locally and how they are connected into the rest of the city.

50 *Rebuilding Urban Complexity*

Box 3   Stockport and its hat cluster

Flemish traders helped to kick-start a hat-making cluster in Stockport from the 15th century onwards, and it continued into the 20th century; indeed, 80 per cent of UK jobs in felt hat production were still in Stockport in 1921 [19]. This is another example of "scale of resolution" becoming important when looking at complexity. While this was a specialised cluster, the construction of felt hats involved over a hundred separate operations, with related "ancillary" industries including machinists, coppersmiths and block makers. The cluster included purpose-built factories, but production also occurred in a multitude of smaller workshops in backyards, particularly on Canal Street, and on surrounding farms as well. Indeed, in earlier times when Stockport was relatively peripheral, the workers often combined hat-making with agricultural work.

Mapping differential economic accessibility

In my own research, I first became fascinated by the way in which different economic sectors position themselves within the street network when looking at historic cities. While researching the city of Antwerp in Flanders I was given access to a recently digitised cadastral map from 1835 and a commercial almanac for 1838, which I used to map the occupations held by the inhabitants of some 10,667 plots [9]. For this city, I found that economic activities were relatively broadly distributed throughout the street system, as opposed to being clustered at particular points (echoing previous space syntax research). However, there were certain trades and occupations which were more frequently found on the city's foreground network of more accessible streets, with this being shown to be statistically significant. The merchants associated with the city's international port had particularly accessible plots at all scales of movement, from local to global. Local artisans, who specialised in sectors such as metals, textiles and furniture, were also found in more spatially accessible streets. In this case these firms would have valued access to the circulation of goods, products and knowledge as much as to the circulation of people.

Employing this methodology for the contemporary city of Greater Manchester produced new challenges, due to the confidentiality of more recent economic data. Nevertheless, it revealed broadly comparable results. In Greater Manchester too, most economic activities are highly dispersed. Indeed, when all the industry sectors which make up the city economy are seen together on a map, it gives an impression of city that is teeming with economic diversity. However, the textiles and clothing industries were more clustered in the centre of the city and just to the north of this centre. Retail, wholesale and knowledge-based services were also more likely to be found in more central areas at the citywide and also local

Configurational analysis of economies and space   51

scale. Perhaps surprisingly, small manufacturing firms were also found in central areas, in both the main city centre and in "satellite" centres such as Bolton and Ashton-under-Lyne [20].

Street-based space syntax analysis revealed how some economic sectors located themselves in streets of "above average" spatial integration, and/or sought higher levels of through movement at a series of scales (2 kilometres, 10 kilometres and 100 kilometres) while others sought more segregated parts of the system. Services based on customer interaction (retail, wholesale, food and hospitality, and real estate) were found, as might be expected, on streets that were local centres but also had higher-than-average through movement. One example of such a location is Castle Street in Edgeley, Stockport. The street is particularly well integrated into its local urban fabric while also being part of the foreground network of the city – a spatial condition further strengthened in the past by its also being on the tram network. A visit to this street identified fifty-five local businesses, specialising in thirty-two different goods and services, among them haberdashery and fabric shops; pubs, bakeries and restaurants; florists and DIY stores; opticians, estate agents, funeral services and banks [20]. While some sectors were overrepresented (with the street offering thirteen different hairdressers and barbers), Castle Street was found to offer a broad cross-section of services and products for local residents, most of which were independent firms, making this a diverse local service hub. It echoed places that Douglas Rae mapped in New Haven, Connecticut, during its economic peak where 'the stores are so close together in some places that the names must be left out of the illustration' [21, p. 85].

Manufacturing firms were found in Greater Manchester – as expected – on less integrated backstreets, not far from the foreground network of streets. In fact, the importance of a network of backstreets supporting commerce and the production and distribution of goods is acknowledged in the naming of streets in the city. Many foreground network streets share a name with a parallel street which incorporates "back" into its name as a prefix – with China Lane running adjacent to Back China Lane in the Northern Quarter, for example (as can be seen in the annotated GOAD map in Figure 7.4, Chapter 7). Commercial plots stretch between these two types of street, providing a back entrance for delivery firms and suppliers.

At the time of my research, knowledge-based services were also particularly likely to be found in both locally and globally integrated streets in Greater Manchester – with the financial services, insurance, legal, science, creative and advertising sectors showing particularly high values of integration at radii of 2 kilometres and 10 kilometres. Again, these sectors would seem to be seeking both local accessibility and global accessibility to the rest of the city, as has been found in other cities such as Barcelona [22].

A space syntax of agglomeration externalities?

One of the reasons why economic activities may be seeking accessibility to the foreground network of the city, and to local integrated centres, is to take advantage of the agglomeration externalities associated with the exchange of capabilities

52   *Rebuilding Urban Complexity*

between firms – including common labour pools, knowledge sharing and supply chain linkages.

As noted earlier, the foreground network of cities plays an important role when it comes to daily commutes, maximising the amount of the city which is accessible within "acceptable" commutes of around 40–60 minutes. It therefore supports the labour matching and labour sharing that is key to how cities work as "functional labour market areas". Angel and Blei [23] identify that as cities grow larger, they "self-adjust" to ensure that labour markets are still accessible within an acceptable commute time, through the provision of effective infrastructure. These authors equate such infrastructure with the rapid expressways and inner-city motorways which now dominate many U.S. cities (and indeed were introduced in British cities from the 1960s onwards). Bill Hillier argued, however, that such adjustment also occurs more organically, based on "deformed wheel" foreground structures [5]. The longer lines associated with the foreground network ensure that people can still rapidly access all parts of the city as the city grows.

In Greater Manchester, not all the surrounding local authorities are equally well connected into the city's foreground network, and this appears to have affected commuting patterns. Wigan, for example, is relatively peripheral to the rest of the city in this network, and has cross-city commuting links that are weaker than those of other local authorities, apart from with its neighbour Bolton (see Figures 4.3 and 5.2).

When it comes to both labour market matching and *knowledge sharing* in the city, local integrated centres are also important. "Griddy" areas that host a large number of intersections increase opportunities for encounter, and hence foster the exchange of information on local job openings and projects that may lead to new forms of innovation – what is commonly known as the "word on the street" [5]. Indeed, local street network density has been found to correlate with local patent applications in a large-scale study in the United States [11].

More integrated local streets also provide the footfall necessary for the "third spaces", such as cafés, pubs, bars and restaurants where ideas are likely to be shared and new economic projects begin. One area which is particularly well set up to support such local knowledge sharing is Altrincham, in the south-west of the city. This is a neighbourhood where the arts, entertainment and recreation and professional, scientific and technical sectors are particularly concentrated. As was identified in Chapter 4, Altrincham is a spatially integrated local centre, and it supports a rich variety of local cafés and third spaces which, as Urry [24] points out, allow knowledge to be exchanged through "playful" forms of sociability. Despite the fact that Altrincham is relatively peripheral, being 13 kilometres from the heart of Manchester, this area also provides multiscale accessibility, being connected back into the city's core through the foreground network of streets, permitting what Read et al. [25] would call the local 'insertion of the metropolitan scale'.

Configurational analysis of economies and space 53

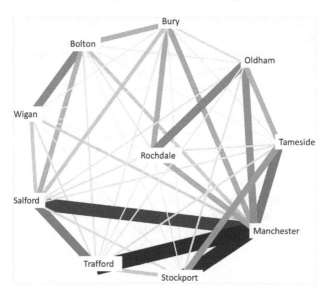

Figure 5.2 Commuting flows between Greater Manchester local authorities

Note: Based on place of work and place of residence from the 2021 Census. This diagram is produced in Gephi using a Fruchterman-Reingold placement algorithm which is slightly different to Force Atlas (used for the network diagrams in Chapter 3)

Source: Author's own diagram

Metropolitan-scale encounters and contacts are likely to be the most important when it comes to more *radical* types of innovation in cities, which are often "recombinant" – that is, based on the *cross-sector* fertilisation of knowledge. Jane Jacobs celebrated cities for providing access to the unexpected, or 'the strange' [26, p. 311], while Hillier and Netto [27] pointed out that while cities create dense local relationships between people, the foreground network of cities also permits the development of citywide social networks which they call the 'large graph'. These latter authors drew on Granovetter's ideas about the importance of 'weak ties' [28] to argue that the most successful city dwellers "globalise" their networks to meet people whom they did not know they needed to meet – contacts of the "right kind" who would be close enough in their thinking to be relevant but far enough away to produce something new [27, p. 197–198].

As an illustration, an area of Manchester which is particularly well placed to support more global knowledge exchange is the Oxford Road corridor. Benefiting from both local integration and citywide connectivity, the area hosts more than 70,000 jobs, over half of which are within knowledge-intensive sectors [29], while accommodating two universities (the University of Manchester and Manchester Metropolitan University), a science park and a number of health care research institutions. Spatially, the area provides a linear extension of the city centre towards

54   Rebuilding Urban Complexity

the south. As the city has expanded, the road has formed an increasingly important part of both the "integration core" of the city and the foreground network of the city. This positioning connects the institutions to the broader knowledge capabilities of the city (in part because, as we will see in Chapter 9, this area managed to escape being closed off into a more inward-looking "knowledge precinct" during the 1960s).

It is important to mention that while knowledge sharing is often seen as a key "agglomeration externality" and mutual benefit acquired by firms in cities, not all firms seek to freely share their information. In fact, knowledge can be a hard-won and precious commodity which it is important to guard rather than share (see Box 4). The Bartlett academic Alan Penn points out that architecture helps to manage the tension between the need for exposure and refuge, through simultaneously opening up and restricting visual and movement-related accessibility [30] – something which is particularly important, given that acquiring information is as much a visual as an oral exercise.

Box 4   Not everybody wants to share their knowledge

Knowledge is often a closely guarded commodity. The anthropologist Pnina Werbner [31] describes how Pakistani entrepreneurs in Greater Manchester in the 1990s gained their knowledge by trial and error over relatively long periods of time, and shared it based only on established trust relationships (and processes of "triangulation", or introductions through somebody who was already trusted). The Alliance Project [32] also reported that many textiles firms in the North West were relatively secretive, guarding knowledge through a mix of formal processes (such as patents) and informal processes (such as feigned complexity– pretending that a mode of production was more complicated than it really was). The latter phenomenon has also been found locally in the engineering sector [33, p. 26].

The built environment also plays an important role in affording people *protection* from encounter. The manager of Xpose, a woollen hat manufacturer interviewed as part of my PhD, was pleased that they were in a location where it was harder for other firms to copy their ideas, particularly through seeing the machines that they were using [20]. Elsewhere in Cheetham and Broughton, firms seek secrecy for very different reasons, such as concerns about illegality and trading standards. The sale of counterfeit goods has long damaged the reputation of the area. Indeed, Andy Spinoza identifies that this is where many counterfeit T-shirts were made to capitalise on the wave of youth culture associated with the "Madchester" dance and drug scene in the 1980s and early 1990s [34].

Urban "vantage points" into global supply chains

While street networks provide an important affordance for labour and knowledge sharing *within* the city, the foreground network of streets is also important in providing a link *outside* the city, supporting product sharing to more geographically distant supply chains. Economic geographers are keen to point to the importance of local-to-global links within supply chains, with the Marxist geographer Doreen Massey being particularly interested in the 'articulations of social relations' which occur across different geographical scales, and Bathelt, Malmberg and Maskell famously highlighting the symbiotic relationship between 'local buzz and global pipelines' [35]. Nevertheless, there is rarely an acknowledgement that the structure of local street systems might help to shape access into these pipelines [1].

In the case of Greater Manchester, there are niches which provide particularly useful "vantage points" or points of access into such networks, due to the configuration of the street system. One such area is Strangeways in the north of the city. As identified earlier, it is rare to find small clusters of specialised industries in the street network in contemporary Greater Manchester. Nevertheless, a very obvious cluster, which I was shown by a local policy maker on one of my first research trips to the city, is the fashion-wholesale cluster which dominates the grid of streets just north of Strangeways prison. This area might seem to be an unlikely place to highlight, given its association with crime and questionable activities in the clothing trade (see Box 4). Nevertheless, Strangeways – and the wider area of Cheetham Hill – has long provided a base for textiles, clothing and tailoring in the city (see, for example, the research carried out by Laura Vaughan on Jewish tailoring in this area in the late 19th century [36]). Today there is an astonishing concentration of fashion wholesalers along the parallel streets of Derby Street and Broughton Street, mostly specialising in relatively inexpensive "fast fashion". At first sight the businesses appear to be retail stores, with bright and attractive shop fronts displaying clothes, bags, shoes and fancy dress. However, most also exhibit "Trade Only" signs. The area hosts a diverse but interconnected range of industries, including six different wholesale sectors, alongside creative, manufacturers and retailers (see Figure 5.3). In addition, a set of suppliers support the cluster, including photographers, shop fitters and delivery agents but also cafés, restaurants and faith institutions – what Jane Jacobs called supporting "secondary diversity" [26].

Strangeways provides an example of the subtle ways in which 'urban areas create a sense of local structure without losing touch with the larger-scale structure of the system' [5, p. 101]. The commercial activities spill out into the larger street system, rather than being enclosed or segregated (as is often found in business or industrial parks). Indeed, the streets are well-knitted into the surrounding urban fabric, and show high levels of integration at a two-kilometre radius. More importantly, however, the area sits between two spokes of the "deformed wheel" foreground structure of the city which connects central Manchester to the north: Bury New Road and Cheetham Hill Road (see Figure 5.4). The foreground network renders Strangeways "shallow", or more accessible to the rest of the United Kingdom,

56 Rebuilding Urban Complexity

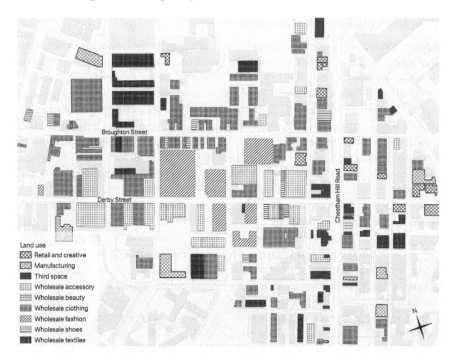

Figure 5.3 Fashion and textiles cluster in Strangeways

Source: Map produced by Nicolas Palominos, based on fieldwork by the author in Strangeways described and mapped in [20]. Base map tiles by CartoDB, under CC BY 3.0. Data by OpenStreetMap, under ODbL

providing an excellent vantage point from which to engage in the national and international trade which underlies the fashion cluster.

The importance of the global reach of the area to local business was made clear through local interviews [20].[2] The haberdashery wholesaler Jay Trim (see Box 8) identified that they send out approximately five hundred parcels a week, with three carriers coming in every day for collections of products. The firm had one hundred suppliers overall, with only 10 per cent of their range being manufactured in the UK. Many of their relationships with suppliers are managed via the internet. This is a relatively new phenomena – in the mid-1970s, 70 per cent of their customers would have been within an eight-kilometre radius. The Strangeways neighbourhood also has local affordances which facilitate the movement and storage of goods, including a side street which has been taken over by a local international deliveries broker, and a rare combination of shop-style frontages onto the streets and deep, receding warehouse spaces behind.

Exploring the spatial arrangement of industry relatedness in a city

It is clear from this analysis that city street systems provide a set of network-based affordances that support the operation of agglomeration externalities and

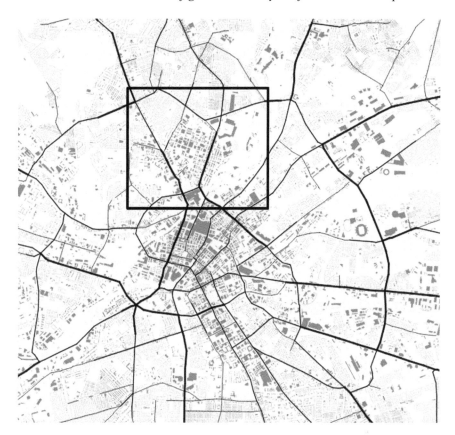

Figure 5.4 Positioning of Strangeways between two spokes of the foreground network
Source: Author's own diagram. Building layer from OS Data © Crown copyright and database rights [2025] Ordnance Survey (100025252) OpenMap – Local

"economic part-whole" relationships – something which is rarely acknowledged in the economics literature. Firms choose different parts of the street network that offer accessibility to both markets and other firms at multiple scales, from the local streets of Atrincham to the international trade networks so important to Strangeways. But what about the topological "industry relatedness" structures set out in Chapter 3? Is it possible to find a spatial underpinning for the skewed or uneven configurational economic relationships which exist between industries, with certain industries being more likely to share labour, knowledge and products? While there has been a good degree of research into how industry relatedness leads to agglomeration at city and regional scale [20, 37, 38], there is much less focus on how related industries are spatially organised *within* cities.

Overall, it seemed from my research that Greater Manchester is acting as 'one giant workshop' [10, 39], with related or economically proximate industries sharing

58 Rebuilding Urban Complexity

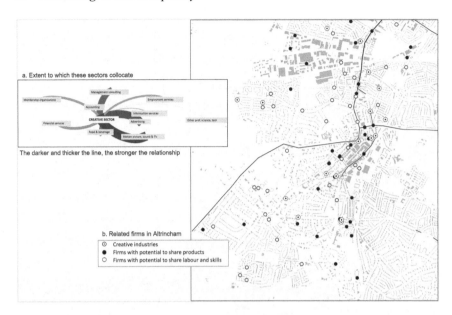

Figure 5.5 Mapping "fields of collaboration potential" in Altrincham

Source: Author's own diagram. Building layer from OS Data © Crown copyright and database rights [2025] Ordnance Survey (100025252) OpenMap – Local. Company sector data was sourced from the FAME database compiled by Bureau van Dijk (now part of Moody's) in 2018

labour pools, knowledge sharing and supply chains across the city as a whole, supported by the foreground network of streets. Statistical analysis also revealed, however that *skills*-related sectors seem to seek more *local* proximity to each other at neighbourhood scale within the city, with skills-related knowledge-based services being particularly likely to collocate this scale [20]. The creative industries were generally found, for example, in the same neighbourhood as other skills-related knowledge-based services – management consulting, advertising and media production (motion picture, sound and TV) (see Figure 5.5). They were also found in proximity to companies with which they are likely to cooperate through supply chains, such as financial and information services. By way of illustration, the map in Figure 5.5 explores the "field of collaboration" open to these creative firms in the area of Altrincham which we have already visited. It shows where creative-sector firms are located in the area, and identifies the location of their "frequent neighbours". Interestingly, those sectors with which the creative sector has closer supply chain relationships are clustered around Altrincham's town centre, while many of the firms with which the creative sectors have an above-average skills overlap are scattered across the residential area, suggesting that these industries often involve working from home.

Such mapping provides an interesting way of visualising what Jacobs called the 'close-grained juxtaposition of talents' [26, p. 584] permitted by cities. Whether

Configurational analysis of economies and space 59

or not firms actually interact with each other, however, will depend on a number of other factors, not least whether local firms have some discretion in local relationship-forming, as opposed to being branches of a larger, centralised operation. Nevertheless, Altrincham clearly combines strong economic and spatial potentials – a cluster of related knowledge-based industries hosted in a locally integrated street system connected into broader city-wide movement flows, with all the possibilities for economic synergy that this brings.

Connecting knowledge across neighbouring cities: the Northern Powerhouse?

At a larger geographical scale, recent research has revealed how the related sets of industries found in one city might prove useful to its neighbouring cities, should transport interconnections improve. This is particularly relevant to the idea that northern cities such as Manchester, Leeds and Sheffield might join forces to create a more regional agglomeration in the form of a "Northern Powerhouse" [40]. Research carried out with complexity theorist Neave O'Clery and colleagues at University College London revealed that Manchester, Sheffield and Leeds host particularly complementary sets of embedded knowledge and industrial capability [41]. Reducing travel times between these cities to 40 minutes would, for example, increase the productivity of a set of Sheffield-based industries including pharmaceuticals, paper manufacturing and media broadcasting. Such interrelated economic potentials are often neglected by people carrying out cost-benefit analysis on infrastructure projects. Indeed, as we will see in Chapter 8, the embedded capabilities of Manchester and Sheffield have already proved complementary, particularly in the production of goods associated with the terrain which divides these two cities in the Peak District.

In summary

In this chapter we have explored how the configurational relationships which structure urban economies and urban street systems work together. We have seen how economic part-whole relationships and spatial part-whole relationships play a mutually reinforcing role within processes of urban complexity. While economic sectors seek accessibility within the street system in different ways, the foreground network brings the whole system together to function as an economic whole. The spatial configuration of the street system supports "agglomeration externalities" at different scales – while also allowing companies in some parts of the city privileged access into international supply chains. This provides a window of understanding into how urban economic systems are 'spatially emergent and spatially embedded' [42] – with the *networked* basis of this embedding being too rarely acknowledged in the economics literature.

In the next part of the book, we will turn to another principal characteristic of complex systems: that of *evolution*. We will consider how the current interdependences which exist in cities have emerged over time, looking at historical processes of economic branching and self-organisation in spatial systems. We

60 *Rebuilding Urban Complexity*

will revisit Greater Manchester from this historical perspective but also explore similar processes in Sheffield and other post-industrial cities. As in Part One, we will first explore economic evolution, then spatial evolution, and then see what can be understood from bringing analysis of these two complex processes together.

Notes

1 Merriam-Webster Dictionary – Online Edition. www.merriam-webster.com. Accessed [07/06/2024].
2 This book includes information from, and profiles of, the Greater Manchester-based companies Jay Trim, Private White V.C., Wright Bower and Xpose, based on interviews carried out during my PhD – I am grateful to each of these companies for their rich and detailed contributions to this research.

References

1. Froy, F., Learning from architectural theory about how cities work as complex and evolving spatial systems. *Cambridge Journal of Regions, Economy and Society*, 2023. **16**(3): p. 495–510.
2. Overman, H.G., S. Gibbons, and A. Tucci, *The case for agglomeration economies.* 2009, Manchester: Manchester Independent Economic Review.
3. Hillier, B., *Between social physics and phenomenology: explorations towards an urban synthesis?* 5th International Space Syntax Symposium. 2005, Delft.
4. Glaeser, E., *Triumph of the city: how urban spaces make us human.* 2011, London: Pan Macmillan.
5. Hillier, B., *Space is the machine: a configurational theory of architecture.* Vol. 2007 e-print. 1999, Cambridge: Cambridge University Press.
6. Penn, A., I. Perdikogianni, and C. Motram, Generation of diversity. In *Designing sustainable cities*, G.E.R. Cooper and C. Boyko, Editors. 2009, Chichester: Wiley Blackwell.
7. Vaughan, L., *Suburban urbanities: suburbs and the life of the high street*, L. Vaughan, Editor. 2015, London: UCL Press.
8. Vaughan, L., et al., *An ecology of the suburban hedgerow, or: how high streets foster diversity over time.* 10th International Space Syntax Symposium. 2015, London.
9. Froy, F., Understanding the spatial organisation of economic activities in early 19th century Antwerp. *The Journal of Space Syntax*, 2016. **6**(2): p. 225–246.
10. Griffiths, S., Manufacturing innovation as spatial culture: Sheffield's cutlery circa 1750–1900. In *Cities and creativity from the renaissance to the present*, I.V. Damme, B. De Munck, and A. Miles, Editors. 2018, New York: Routledge Advances in Urban History. p. 127–153.
11. Roche, M., Taking innovation to the streets: micro-geography, physical structure and innovation. *Review of Economics and Statistics*, 2020. **102**(5): p. 912–928.
12. Parham, E., S. Law, and L. Versluis, *National scale modelling to test UK population growth and infrastructure scenarios.* 11th International Space Syntax Symposium. 2017, Lisbon.
13. Hanna, S., J. Serras, and T. Varoudis, *Measuring the structure of global transportation networks.* 9th International Space Syntax Symposium. 2013, Seoul.
14. Law, S., et al., *The economic value of spatial network accessibility for UK cities: A comparative analysis using the hedonic price approach.* 11th International Space Syntax Symposium. 2017, Lisbon.

Configurational analysis of economies and space   61

15. Narvaez, L., A. Penn, and S. Griffiths, The spatial dimensions of trade: from the geography of uses to the architecture of local economies. *ITU Journal of the Faculty of Architecture*, 2014. **11**: p. 209–230.
16. Scott, F., High street productivity. In *Suburban urbanities: suburbs and the life of the high street*, L. Vaughan, Editor. 2015, London: UCL Press.
17. Hanson, J., *Order and structure in urban space: a morphological history of the City of London*. Bartlett School for Architecture and Planning. 1989, London: University of London.
18. McKnight, P., *Stockport hatting*. 2000, Stockport: Stockport MBC, Community Services Division.
19. Clay, H. and K.R. Brady, *Manchester at work: a survey*. 1929, Manchester: The Manchester Civic Week Committee.
20. Froy, F., *'A marvellous order': how spatial and economic configurations interact to produce agglomeration economies in Greater Manchester*. Bartlett School of Architecture. 2021, London: University of London.
21. Rae, D.W., *City: urbanism and its end*. 2005, New Haven: Yale University Press.
22. Muniz, I. and M.-A. Garcia-Lopez, The polycentric knowledge economy in Barcelona. *Urban Geography*, 2010. **31**(6): p. 774–799.
23. Angel, S. and A. Blei, *Commuting and the productivity of American cities: how self-adjusting commuting patterns sustain the productive advantage of larger metropolitan labour markets*. Working Paper 19. 2015, Brooklyn: Marron Institute of Urban Management.
24. Urry, J., *Mobilities*. 2007, Cambridge, UK: Polity Press.
25. Read, S., et al., *Constructing metropolitan landscapes of actuality and potentiality*. 6th International Space Syntax Symposium. 2007, Istanbul.
26. Jacobs, J., *The death and life of great American cities [1993 edition]*. 1961, New York: The Modern Library.
27. Hillier, B. and V. Netto, Society seen through the prism of space: outline of a theory of society and space. *Urban Design International*, 2002: p. 1–22.
28. Granovetter, M., The strength of weak ties. In *Social structure and network analysis*, P.V. Marsden and N. Lin, Editors. 1982, Beverly Hills: Sage Publications Inc. p. 101–130.
29. Manchester City Council, *Oxford road corridor strategic regeneration framework guidance*. Report for Resolution. 2018, Manchester: Manchester City Council.
30. Penn, A., The city is the map: exosomatic memory, shared cognition and a possible mechanism to account for social evolution. *Built Environment*, 2018. **44**: p. 162–176.
31. Werbner, P., Renewing an industrial past: British Pakistani entrepreneurship in Manchester. In *Migration: the Asian experience*, J.M. Brown and R. Foot, Editors. 1994, Basingstoke: Palgrave Macmillan.
32. Alliance Project, *National textiles growth programme: supply chain mapping and targeting the investment pipeline*. 2016, London: Alliance Project.
33. Volterra, *Innovation, trade and connectivity*. 2009. https://www.greatermanchester-ca.gov.uk/media/6669/mier-tradeinnovations.pdf.
34. Spinoza, A., *Manchester Unspun: pop, property and power in the original modern city*. 2023, Manchester: Manchester University Press.
35. Bathelt, H., A. Malmberg, and P. Maskell, Clusters and knowledge: local buzz, global pipelines and the process of knowledge creation. *Progress in Human Geography*, 2004. **28**(1): p. 31–56.
36. Vaughan, L. and A. Penn, Jewish immigrant settlement patterns in Manchester and Leeds 1881. *Urban Studies*, 2006. **43**(3): p. 653–671.
37. Diodato, D., F. Neffke, and N. O'Clery, Why do industries coagglomerate? How Marshallian externalities differ by industry and have evolved over time. *Journal of Urban Economics*, 2018. **106**: p. 1–26.
38. Ellison, G., E.L. Glaeser, and W.R. Kerr, What causes industry agglomeration? Evidence from coagglomeration patterns. *American Economic Review*, 2010. **100**: p. 1195–1213.

39. Griffiths, S. *Persistence and change in the spatio-temporal description of Sheffield parish c.1750–1905.* 7th International Space Syntax Symposium. 2009, Stockholm.
40. Lee, N., Powerhouse of cards? Understanding the 'Northern powerhouse'. *Regional Studies*, 2017. **51**(3): p. 478–489.
41. Straulino, D., et al., Connecting up embedded knowledge across Northern powerhouse cities. *Environment and Planning A: Economy and Space*, 2023. **55**(7). https://doi.org/10.1177/0308518X231159108.
42. Martin, R. and P. Sunley, Complexity thinking and evolutionary economic geography. *Journal of Economic Geography*, 2007. **7**(5): p. 573–601.

Part Two

6 How local economies evolve and branch

The network theorist Barabási argued that 'to explain a system's topology we first need to describe how it came into being' [1, p. 413]. In this second part of the book, we will explore how economic and spatial configurations evolve, beginning in this chapter with *economic* evolution and branching.

Economic branching and the dynamic division of labour

Systems rarely stand still. They evolve and change over time. The evolution of urban economies was a particular focus of Jane Jacobs' lesser-known book *The Economy of Cities* [2], where she argued that the division of labour in cities can be largely stagnant (as in a factory which divides tasks into separate production lines in the drive for efficiency) or it can be creative – with economic innovation leading to the continual development of new "footholds" for diversification. She described a "branching" process, where new work emerges from older work (or in fact, fragments of older work), as people solve problems based on the materials and technologies that they already have to hand. She argued that this underlies the very origins of cities, with local trading posts becoming more important through the gradual multiplication of different types of economic activity. In the process, exports led to new imports, which were then imitated and reproduced locally, leading to new types of exports in an ever-expanding process. As she writes, 'the greater the sheer numbers and varieties of divisions of labor already achieved in an economy, the greater the economy's inherent capacity for adding still more kinds of goods and services' [2, p. 59]. Jacobs also noted that cities are from the start partially open systems, due to their embedding in trading networks. The idea that economic diversity begets greater economic diversity has been substantiated more recently, in a study of 366 metropolitan statistical areas in the United States where economic diversity was found to increase in a super-linear (or exponential) way over time, whereas the total number of establishments and employees only increased "linearly" as cities grew [3].[1]

As was often the case, Jacobs was influenced in her writings on urban economies by wider complex systems theory, including the analysis of biological and ecological systems. The process by which diversity begets more diversity is in fact common to many different systems: diversity is thus both an input and

DOI: 10.4324/9781003349990-9

66 *Rebuilding Urban Complexity*

an output of evolutionary processes. The information already embedded in a system allows an increased harnessing of energy to build still more structure and information, with each element providing the context or niche for a new element [4, 5].

Complexity theorist Stuart Kauffman also sought to apply biological ideas to economic evolution, describing how each new invention generates complementary niches. He stressed that economic growth largely depends on the development of products and sectors that are *complements* to what has come before, rather than substitutes [5]. Both he and Jacobs felt that the branching process has a very material underpinning. While Jacobs argued that the material properties of tools and materials "suggested" new solutions to problem-solving entrepreneurs, Kauffman similarly describes the "preadaptations" which exist in materials and tools – the affordances of an object which are not yet in use but which could prove useful in different ways in the future. This way of thinking also shows similarities with that of the economist Joseph Schumpeter, who earlier in the 20th century explored the importance of economic evolution to the generation of new innovation and hence to business cycles. Schumpeter argued that evolution is not necessarily a steady and incremental process but occurs in spurts and waves. These continue until all associated possibilities associated with a new innovation are exhausted, and until competitors are eliminated in a process that he called 'creative destruction' [6].

All these ideas have influenced the discipline of evolutionary economic geography and the work on "related variety" and "industry relatedness" which has already been discussed in Chapter 3. Branching processes are, in particular, felt to be the origin of the patterns of related variety or industry relatedness which exist in urban economies. They are also key to the evolutionary path *inter*dependency which is common to cities and regions [see 7–9]. Given the importance of co-evolution of different industry sectors, it is particularly important to consider the *synergies* which exist between sectors, defined by the *Oxford English Dictionary* as 'any interaction or cooperation which is mutually reinforcing: a dynamic, productive or profitable affinity, association or link'.[2] As we will see, synergies have played a key role in the ongoing development of Greater Manchester and other similar cities.

Greater Manchester: a 'city remaking itself out of its own history'[3]

In her BBC Radio series on Manchester, titled 'Alchemical City', the writer Jeanette Winterson described how the city was always 'remaking itself out of its own history' [10]. The city has indeed shown important path (inter)dependencies in its economic evolution. Related sectors such as textiles and chemicals have developed and branched in symbiotic ways, with these relationships being uncoverable through what is known as "ancestry analysis" – a process similar to the tracing of the phylogenetic background of a natural ecosystem (see Figure 6.1). While ancestry analysis can be a very precise and painstaking task, in this chapter, a

How local economies evolve and branch 67

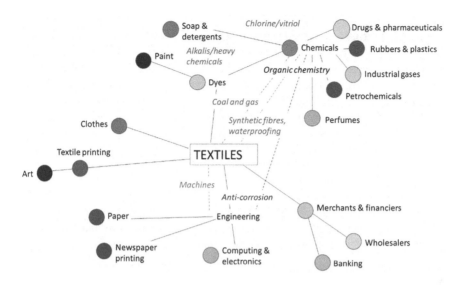

Figure 6.1 Historical economic branching in Greater Manchester. Text in *italics* shows common processes between the sectors

Source: Author's own diagram

broad-brush approach is taken to understanding the synergies underlying the city's evolution, based on archival research.

Textiles

The sector most associated with Manchester and its surrounding towns is textiles (see Box 5). Textiles have been important to the local economy since well before the industrial revolution, with the city hosting the manufacture of woollens in 13th century. In 1772, when the earliest trade directory for Manchester was produced by Elizabeth Raffald, textile trades formed slightly less than one-third of all entries, with fustian cloth (which combined cotton and linen) being a particular local specialisation [11, 12]. The city became best-known, however, for its weaving of cotton fabrics. By the 19th century Manchester and its surrounding Lancashire mill towns were responsible for spinning 32 per cent of *global* cotton production [13]. The development of mass-produced woven cotton fabrics in the city was in fact a classic case of Jacobs-style "import replacement", with local textile producers finding new ways to mechanically imitate the handwoven colour-fast fabrics that had been imported to Britain from India beginning in the 1630s. Indeed, after the 1820s these fabrics were later sold back to India, undercutting the Indian market and provoking Gandhi's famous call for people to return to handmade textile production during the Indian independence movement. The production of clothing in Greater Manchester became more important later, with John Rylands pushing for the diversification into garments from the 1880s onwards, helped by the development of the sewing machine.[4]

68 *Rebuilding Urban Complexity*

Box 5   Textiles – a network-based material

Textiles are interesting in a more abstract sense for this book because they themselves have a systemic quality – mostly being networks made of fibre which has been spun into yarns or threads. Fibre can be natural (wool and cotton) or synthetically produced (acrylics, nylon and polyester), with both natural and human-made fibres being made of *polymers*. These polymers are also interlinked entities, being comprised of long sequences of large molecules linked together by covalent bonds [14] – the term "polymer" derives from the Greek words for "many" and "parts". Fibre can be constructed by weaving, knitting and crocheting, with each leading to different structural capacities in the resulting fabric. Nylon in fact represents the extraction *of the chemical boundary* between two reacting chemicals, providing an apt metaphor for the focus in this book on synergy and interdependency.

Greater Manchester's textile industry was already diverse in the late 19th century, hosting 'bleachers, dyers, calenderers, printers, finishers, sizers, stretchers, embossers, perchers, moranders, winders, warpers and many more' [15, p. 9]. Today this sector is still more diverse, ranging from clothing to homewares and medical textiles and to composite materials for the aeronautics and car industries.

While the textiles industry evolved and diversified on its own terms, it also did so through synergies and interlinkages with a wider range of other local industries. Two tightly coevolving sectors were the chemicals and engineering sectors (and the textiles sector maintains strong links with these today). These industries largely grew up in the city to support the textiles industry but later diversified, and then came to strongly influence the development of the textiles industry itself. Even back in the 1920s, a journalist writing in the *Manchester Guardian* identified that the textiles, chemicals and engineering sectors were an example of 'industrial interdependence' [16].

Chemicals

The close interrelationship between textiles and chemicals began with "finishing" processes – and in particular the bleaching of garments, which first happened in fields around Bolton, where local farmers "tented" textiles in the sun after boiling or "bowking" them in a solution of alkaline plant ash, and then neutralising or "souring" them with milk – with the entire process taking up to six months. The introduction of new chemicals from Scotland, such as sulphuric acid and chlorine-based bleaches, significantly speeded up this process, and Bolton became a significant chemicals town by 1840 [16, 17]. Another branch of the developing chemical industry centred around alkalis, linked to salt extraction in neighbouring Merseyside. This supported the expansion of local soap and glass industries, which remain a specialism today (associated with large firms like Cussons, which make the brand Imperial Leather). A group of alkali-based manufacturers later came together to form ICI (the Imperial Chemical Industries), which become the largest chemical manufacturer in Britain, itself branching into pharmaceuticals, plastics, paints and metals.

A dynamic relationship between the chemicals and energy industries (which today still fall within the same skills basin) had important repercussions for the textiles industry. Alkalis provided the basis of a large dyestuffs and pigments industry, which fed back into textiles development. The industrial by-product coal tar was also used in dye development, a process that first developed in London and then blossomed in the Manchester region [17]. Later still, the petrochemicals companies which clustered along the Manchester Ship Canal fed into synthetic dye production, leading to an explosion of colour possibilities.[5]

The calico printing trade also became important to the city, after having transferred from London to Lancashire in the 1760s and 1770s [18]. Before these printing processes took hold, the only way that people could achieve "design" in textiles and clothing was through the process of making the fabric itself. Now design could be imposed at the end of the process. Again the printing process brought the chemicals and textiles industries together, with one commentator arguing that 'in every large-print work there is either a partner or a manager thoroughly versed in practical chemistry', (quoted in [19, p. xxvii]).

Later the chemicals industries become central not only to the *finishing* of textiles but also to their production, as artificial polymers were created using hydrocarbons from the petrochemicals industry to make new synthetic materials such as nylon, polyester and acrylics. The previously cited *Guardian* journalist remarked that the textiles being produced in the city in the 1920s were as much a product of the chemicals industry as the textiles industry. The era's atmosphere of innovation is depicted in Ealing Studios' 1951 film *The Man in the White Suit*, in which Alec Guinness plays an ambitious worker in a textiles factory who experiments (unsuccessfully) on the side with making a new dirt-resistant fabric in bubbling and steaming systems of test tubes. By the 1980s, 327 million yards of synthetic materials were produced in Lancashire, almost matching the 399 million yards of natural textiles [20]. Interestingly, these new types of fabrics were often produced by more traditional firms which had branched into using the new materials. Traditional mills ended up processing both cotton and rayon, with synthetic fibres in general being handled by existing spinners, weavers and finishers [21]. The branching process which happens at the scale of whole economies also occurs at the scale of firms [see also 22].

Engineering

A second industry which has always had a close relationship with textiles is engineering. Today, Greater Manchester hosts many niche engineering specialisms (at 4-digit classification scale), borne of a long history of this trade in the city. The engineering sector was crucial to supporting and servicing the textiles industry,

70 Rebuilding Urban Complexity

most famously developing the steam engine power which powered the mills. However, it diversified into a wide variety of different products, including 'hydraulic pumps, cranes, fire engines, gears and pumps, washing and laundry equipment, weighing machines, factory clocks, screws and bolts, turnstiles, lathes, and even mangles' [15, p. 9]. Engineering firms often took over former mill buildings and weaving sheds as they developed new lines of work. A particular specialism was machine tools, which were developed by Joseph Whitworth, who would later bequeath his resulting fortune back to the city's art gallery and the Christie Hospital [23]. A relationship to the chemicals industry was also important here, as firms experimented with ways of preventing corrosion [16], and indeed engineering was also an important third player alongside textiles and chemicals in the calico printing industry. By 1921, metalworking and engineering had become the biggest employers after textiles [24]. During the period between the First and Second World War, electrical engineering flourished in the city, with electrical manufacturers Westinghouse (later Metrovicks) and Ferranti diversifying into electronics, domestic appliances and avionics [25]. The city maintains a concentration of electronic component firms today, alongside other enduring specialisms such as machinery for paper, rubber and plastics production; lifting equipment; metal fabrication; and metal treatments.

An evolving open system

It is notable that throughout this process Manchester was continuously open to outside influences.[6] In Elizabethan times, the city was particularly accessible to outside innovation and influence because it did not have a strong guild system. A visitor to the city in 1783 described how 'trade has been kept open to strangers of every description who contribute to its improvement by their ingenuity' [cited in 26, p. 4]. Later, many of the inventions that were key to the development of different aspects of the chemicals, engineering and textiles trade happened outside the city, and indeed outside the country. Clay and Brady asserted in 1929 that 'no city has wider connexions with the rest of the world than Manchester' [24, p. 1].

Merchants and wholesalers based themselves in the commercial heart of the city, with warehouses far outnumbering factories. There were numerous "foreign houses" that linked local manufacturers into diverse international markets. The merchants have been described as the 'pivot around which much of the region's economy evolved' [25, p. 49], while Alfred Marshall himself evocatively described the "foreign houses" as being the means by which different countries around the world 'affixed commercial tentacles to the metropolis of the Lancashire cotton industry' [27, p. 286]. As a result, the warehouses of the city were as important as its factories, if not more. Brian Groom recently described Manchester at this time as 'a warehouse town' [23], with most cotton spinning happening in its surrounding town centres such as Bolton and Oldham. The growth of the warehouses was rapid. In 1820, 126 warehouses were listed in the Township Rate Books. Nine years later, there were nearly 1,000. Before warehouses became their own class of architecture,

How local economies evolve and branch 71

the early warehouses were often converted dwellings or cellars, marked by decorative palazzo styles with 'grand staircases and extensive displays of goods' [19, p. xxvi].

The merchants provided a key link in local production and supply chains: in the textiles industry, "putting out" (sending out yarn to be made into cloth, and then retrieving the completed cloth for finishing), storing, packaging, marketing and distributing. The importance of merchants to manufacturing industries persisted well into the 20th century, with the merchant being described at this time as an 'architect of fabrics' [28, p. 36]. The wholesale and manufacturing industries in Greater Manchester can therefore be seen as tightly interlinked, and tightly co-evolving. The merchants also gained ideas for new fabrics from their interaction with customers from around the globe – an example being the exploitation of Indonesian batik designs to sell fabrics to a burgeoning market on West Africa's Gold Coast [29].

This global interconnectedness had a more disturbing side. Greater Manchester's textile history was closely bound to the slave trade, which was integral to the cotton production which fed the city's textiles factories, coming in through the neighbouring city of Liverpool. Later, as we will see in Chapter 8, the city became dependent on rubber, which was extracted through brutal means from the Congo during the Belgian colonial era, and from English colonies in South East Asia [30]. The city was therefore dependent on global "geographies of extraction"[7] which were shaped by extreme forms of inequality and exploitation. In the case of textiles this dependency ultimately led to a radical response. Manchester played an important role in the abolition movement, ultimately helping to end slavery in the cotton industry through an embargo on cotton imports from southern American states at the expense of the city's own production [31].

Another key outside force which was important in the evolution of the city's economies was immigration. Immigrants first came to the city principally from Ireland and eastern Europe (the latter mainly Jewish migrants fleeing persecution), and later from the Caribbean, South and East Africa, China, India and Pakistan to take up roles in the different industries which required more and more workers as they branched and expanded [see e.g. 32]. This has created an important legacy of ethnic diversity in the city.

A city of many threads

This ancestral tracing has focused on only three main industries crucial to Greater Manchester's path interdependency. Nevertheless, many other local specialisms developed similarly as "threads" of production throughout its history – both related and unrelated to the textiles industry. They include paper (with other textiles towns such as Leeds and Leicester also having this specialism), insurance (which originated in fire insurance for the textiles factories), publishing (with the *Manchester Guardian* being initially set up by textile traders), information technology (with the city developing the first computer to store and run programs), and art (Manchester School of Art was an early product of textile printing).

72   *Rebuilding Urban Complexity*

Many people now also associate the city with sport (particularly football) and music. David Haslam noted that it is impossible to talk about music without mentioning Manchester, nor Manchester without mentioning music [33]. Indeed, a key dimension of the city's "reinvention" in recent decades has been the creative and broadcasting industries. The Manchester pop music scene associated with the Hacienda, Factory Records and bands such as Joy Division and later the Smiths, the Stone Roses, Happy Mondays, Simply Red and Oasis was recognised and exploited by city authorities as a source of growth and job creation [34]. As we will see, the emerging music scene was very much self-organised, taking advantage of the left-over buildings not only of the textiles industry, but also of the failed urban regeneration projects which took place in the 1960s and 1970s.

As noted in the introduction, while many see Greater Manchester as a "post-industrial" city, materials remain an important specialism, which finds echoes within the city's university sector. The University of Manchester's Department of Materials has been identified as one of the largest materials departments in any European university [35], and the city also hosts the Henry Royce Institute (a national institute for advanced materials research and innovation) and the National Graphene Institute. Graphene, the focus of the last-named institute, is a new two-dimensional material consisting of a single layer of carbon atoms, arranged in a honeycomb lattice, which combines strength with flexibility and conducts energy and heat very well. There is widespread excitement about graphene, given that it has the power to act as a new general-purpose technology, with implications for many different industries, including new green technologies. The two researchers who discovered it at the University of Manchester – Kostya Novoselov and Andre Geim – are from Russia, an example of knowledge "reaching into the city" but taking advantage of local embedded capabilities and facilities.

Tracing the economic branching of other cities

Greater Manchester is of course not unique in having a rich history of co-evolving industries. When the economic history of British cities is examined, each has a very particular story to tell. Just across the Pennines, the city of Sheffield, for example, has followed a very different trajectory, with a focus on the metal trades. At one point Sheffield was known as "steel city". As already noted, Alfred Marshall was fascinated by the cutlery trades, which had dominated Sheffield since at least the 13th century [36]. By the late 18th century, the focus was not only on knives and cutlery but also on 'scissors, shears, sickles and arrows' [37, p. 13]. The metalworking cluster was able to operate in a networked and nonhierarchical way, further enabled by a bill which was passed in the House of Commons in 1624 to support the establishment of a new Company of Cutlers which was to admit a 'communality' of 2,000 master craftsmen [37]. The cluster gradually evolved into a broad set of light steel–related trades, and then heavier steelmaking and engineering starting in the 1850s, which was particularly concentrated in the Lower Don Valley.

Sheffield's metalworking past continues to make an imprint today. Potter and Watts [38] identify, for example, how Sheffield's history of cutlery-related

How local economies evolve and branch   73

production has resulted in a cluster of plants specialising in technologically-related metal industries. In 2008, 8,000 employers still worked in fabricated metal products – in metal forging, machine tools and domestic tools, with a small element still making cutlery. Ferrous metals, another focus of local invention and development, have also remained concentrated in the area, while diversification has occurred into related technological fields such as tool steel and high-tensile steel.

Other industrial cities had rather different trajectories. In the North East, Newcastle first developed as a coal mining centre and then transitioned towards shipbuilding, alongside many other associated lines of evolution including salt and glass (with the city, like Manchester, having an important alkali industry) [39]. In Leeds, textiles have also long been important, in this case wool and processes of knitting rather than weaving. The port of Bristol, in the South West, was mainly focused on global trade, and the city played a key role in the slave trade, provoking protest in recent years.

From resilience to decline

While economic evolution relies on processes of self-organisation and emergence, evidently, economic sectors do not just evolve on their own – there is a degree of human agency. As we have already seen in Jane Jacobs' writings, one way in which human beings "evolve" sectors is through *problem-solving* with the materials and technologies that they have in hand, both to resolve existing problems and to access new markets.

In Greater Manchester, the development of dyestuffs and chemicals, for example, resulted from an ongoing process where 'experiment would appear to have at least an equal value with the much more widely advocated one of "produce"' [40]. This leads some to see the development of Greater Manchester's economy as a history of bottlenecks that were resolved through human ingenuity.[8] Sheffield's evolution can also be traced as a process of ongoing problem solving, with the discovery of "breakthrough processes" which then unleashed powerful growth. The type of steel that the city became well known for – crucible steel – was developed by a local clockmaker who had become dissatisfied with the quality of steel available for his clock springs in the 18th century in Doncaster. After a period of experimentation, he became a full-time steel-maker in 1750, with crucible steel coming to dominate production, until new steels (such as stainless steel) arrived in the early 20th century.

One factor which particularly interests evolutionary economic geographers is how urban economic systems not only evolve but also maintain a degree of stability over time, including being resilient to outside shocks [41]. As was already touched on in Chapter 2, one definition of resilience is the capacity of a system to experience perturbations and nevertheless return to the same structure and functions, albeit with some degree of adaptation [42]. There is a significant amount of research into the social institutions and routines which help to create resilience over time through both providing stability and supporting a joint response to change [see, e.g., 43]. In the field of architecture, the UCL academic Sophia Psarra

74 *Rebuilding Urban Complexity*

explores the 'potent synthesis of stability and creativity' which was associated with the city of Venice as it evolved over time, due to the mix of urban form, governance and 'the prevailing institutions of Venetian life' [44, p. 129]. Institutions can be both informal (including social codes and conventions) or more formal (such as research institutions). In Manchester, the long-standing Shirley Institute played a key role in maintaining a research focus in the textiles industry and then promoting new textiles innovation [21] – and there were a raft of other industrial associations, such as the Society of Dyers and Colourists, which acted to both link up different economic actors and provide continuity over time.

The relatedness of industries in a city can have both a positive and a negative impact on resilience. In some cases, having related industries can help cities to recover after economic crises by providing people with alternative employment paths (as discussed at the end of Chapter 3). Nevertheless, relying on a cluster of related industries can also in some cases create "lock-in". Economic geographers have long been interested in the way in which path (inter)dependency can make economies slower to adapt to new economic opportunities [45]. Relatedness can also lead to what in biology are called "extinction cascades" [42]. During such cascades, decline in one sector has knock-on negative effects on its surrounding related industries, meaning that a greater part of the economy is affected.

Retraction and loss of complexity

Indeed, in the case of many of the "post-industrial" cities which are the specific focus of this book, economic evolution has not simply been a story of growth and stability. In the cities of Greater Manchester, Sheffield, Newcastle and elsewhere there have also been periods of turbulence and, ultimately, long-term decline. Greater Manchester's decline began after the Second World War and continued into the 1980s, with a 22 per cent decline in jobs between 1951 and 1981. Engineering and electrical goods jobs nearly halved, while textile-related employment fell by 86 per cent. Despite growth in knowledge-based services, Manchester still had 90,000 fewer jobs in 2013 than it did in 1951 [13].

This decline has evidently resulted in a loss of economic complexity. In comparison with the "creative destruction" that follows waves of innovation, this is rather a form of entropy – a progressive loss of information and structure in certain parts of the economy, particularly manufacturing. This has in part been due to external factors. Competition from cheaper manufacturing in other parts of the world from the mid-20th century onwards led to a progressive "thinning out" of industry in cities and a consequent "unravelling" of complex economic relationships and ecologies of production [46, 47]. The manager of the leather bag–making company Wright Bower asserted, for example, that the textiles industry had become atomised and fragmented, leaving isolated "survivor" firms. This point was reiterated by the manager of Private White V.C., whose company's advertising slogan 'Britain's bravest manufacturer' partly referred to the way in which this business has managed to "go it alone" (see Box 7). While not all industries are in decline, the remaining firms in Greater Manchester are often relatively capital-intensive

How local economies evolve and branch   75

(such as in carpet manufacturing) or else involve networks of smaller firms that take advantage of cheap spaces and the flexibility and support of ethnic ties [47].

In Sheffield too, competition for all branches of metal-making, including the cutlery industry, hit hard, with employment in the Lower Don Valley falling from 40,000 to 13,000 in just 10 years between the 1970s and 1980s. In both Sheffield and Greater Manchester, the trade-marks which once celebrated and protected local capabilities have been sold abroad, while machines (those very material embodiments of capabilities) have also left these cities following a slackening of protective legislation in the UK in the 1980s. The loss of embedded local manufacturing ecologies is a poignant reminder that the complex capabilities which make up intertwined economic networks will remain only as long as people are working in these fields. As Hillier and Hanson [48, p. 206] stated, '[W]ithout embodiment and re-embodiment in spatio-temporal reality, structure fades away'.

However, while external economic waves of change undoubtedly play a role in such processes of decline, internal systemic factors also seem to play a role in the progressive loss of economic complexity. In Greater Manchester, for example, an important factor in the decline of the city was the progressive loss of diversity which accompanied the consolidation of companies into fewer and fewer large textile companies. This is a factor which led Jane Jacobs to despair of the city due to its being an "overly efficient company town". She compared the city unfavourably to the neighbouring city of Birmingham, saying that

> Birmingham's economy has remained alive and has kept up to date. Manchester's has not. Was Manchester, then, really efficient? It was indeed efficient and Birmingham was not. Manchester had acquired the efficiency of a company town. Birmingham had retained something different: a high rate of development work.
>
> [2, p. 89]

Jacobs [49] pointed out that cities can sow the seeds of their own destruction when they become too rigid in the way that they host production. The evolutionary economist Stanley Metcalfe [50, p. 64] similarly noted that competition between firms can lead to a process in which 'evolution consumes its own fuel', destroying the variety from which it grew. In Greater Manchester, this reduction in variety was also supported by government legislation, such as the Cotton Industry Act of 1959, which consolidated the textiles industry into the hands of several large-scale players, who resisted further diversification and were less adaptive to change [51]. Government schemes to attract industry away from centres of growth like Manchester to struggling UK regions also played a role in the decline through 'Industrial Development Certificates' [51] – with the policy makers of the time having little understanding that agglomeration economies too need continual reinvestment. The loss of industry in Greater Manchester did, however, create new "lines of flight" [52]. Jacobs wrote '[D]ull, inert cities, it is true, do contain the seeds of their own destruction and little else. But lively, diverse, intense cities contain the seeds of their own regeneration' [49, p. 585]. In Manchester's case such a "seed"

76   *Rebuilding Urban Complexity*

was the already-mentioned contemporary music industry, fuelled by the context of disappointment, dysfunction and discord which existed in the city in the 1970s. As the former Manchester DJ David Haslem writes, '[O]ut of the trauma of the city comes energy' [53, p. xxx].

A fossil capital?

The ongoing economic branching of cities such as Greater Manchester is not only powered by human agency and energy, however. It is all too easy to forget that these processes of bottom-up economic evolution and diversification can occur only when there is also a very material *energy* source which fuels them [54]. In his book *Fossil Capital*, Andreas Malm underlines the importance of coal to the industrial revolution, and stresses that the longstanding processes of economic evolution in Greater Manchester and other industrial cities have led not just to prosperity but also to climate change. He points out that 'only now is it becoming apparent what it really meant to burn coal and send forth smoke from a stack in Manchester in 1842' [55, pp. 4–5]. Indeed, he considers climate change to be a profoundly historical and path-dependent process, and suggests that we revisit such emergent urban histories with our 'eyes wide open'.

In summary

In this chapter we have explored how the economic structure of a contemporary city can in part be explained through branching diversification processes throughout history. We have stressed the importance of path *inter*dependency in building the interconnected economic capabilities so important to today's cities. We have touched on the processes of resilience, decline and then reinvention which have characterised post-industrial urban economies such as those of Greater Manchester, Sheffield and Newcastle. We have also touched on the negative aspects of how cities operate as "partially open" and materially embedded systems – highlighting the historic entanglement of cities such as Manchester in processes of global exploitation and colonialism, in addition to the longer-term unintended consequences of an economy based on fossil fuels. We will return to the last-mentioned concerns later in the book, while also revealing some of the more positive environmental processes which occurred in Victorian cities – particularly around the reuse of waste. Firstly, however, the book will again "shift gear" to look at the emergent processes important to the development and evolution of urban public space and urban street systems.

Notes

1 Jeffrey Lin in 'Cities, innovation and new work' (2007) also found that new types of work and occupation were more likely to be found in U.S. cities that hosted industrial variety, in addition to high numbers of college graduates.
2 Oxford English Dictionary – Online Edition. www.OED.com. Accessed [07/06/2024].
3 BBC Radio 4, Friday, 12 December 2014. https://learningonscreen.ac.uk/ondemand/. Accessed [07/06/2024].

How local economies evolve and branch   77

4 The sewing machine was invented in parallel in the 1850s in Oldham and in Boston in the United States. Oldham became a centre for sewing machine production, while the founding factory Bradbury & Co. went on to also produce motorcycles. See www.sewmuse. co.uk/bradbury/bradburyindex.htm. Accessed [07/07/2024].
5 Unfortunately, Manchester was forced to compete with another successful agglomeration, on the banks of the Rhine in Germany, which also hosted tightly interwoven production processes created by 'much accumulated know-how' (see Desrochers, P., Bringing inter-regional linkages back in: industrial symbiosis, international trade and the emergence of the synthetic dyes industry in the late 19th century. *Progress in Industrial Ecology: An International Journal*, 2008. **5**(5–6): p. 465–481). Nevertheless, local firms like Clayton Aniline Co. and Levensteins became important suppliers of dyes and chemicals for local cotton manufacturers and finishers, and were aided during the First World War as associated trade barriers led to a process of "import substitution".
6 Interestingly, Tim Edensor (2010) identified the complex international relationships which have been important to Greater Manchester's history by focusing on a very material aspect of the city in 'Building stone in Manchester' in *Re-shaping cities: how global mobility tranforms architecture and urban form*, M. Guggenheim and O. Soderstrom, Editors. 2010, Abingdon, Oxon: Routledge.
7 See https://architizer.com/blog/inspiration/stories/saskia-sassen-geographies-of-extraction/. Accessed [07/06/2024].
8 With thanks to Richard Horrocks from the University of Bolton for a rich and inspiring conversation on the history of the chemicals and textiles industries in Greater Manchester.

References

1. Barabási, A.-L., Scale-free networks: a decade and beyond. *Science*, 2009. **325**(5939): p. 412–413.
2. Jacobs, J., *The economy of cities*. 1970, New York: Vintage Books.
3. Youn, H., et al., Scaling and universality in urban economic diversification. *Journal of the Royal Society Interface*, 2016. **13**.
4. Kay, J.J., An introduction to systems thinking. In *The ecosystem approach: complexity, uncertainty, and managing for sustainability*, D. Waltner-Toews, J.J. Kay and N.E. Lister, Editors. Chichester: Columbia University Press, p. 3–13.
5. Kauffman, S.A., *Reinventing the sacred: a new view of science, reason and religion*. 2008, New York: Basic Books.
6. Schumpeter, J., *The theory of economic development*. 1934, Boston, MA: President and Friends of Harvard College.
7. Frenken, K., F. Van Oort, and T. Verburg, Related variety, unrelated variety and regional economic growth. *Regional Studies*, 2007. **41**(5): p. 685–697.
8. Boschma, R. and S. Iammarino, *Related variety and regional growth in Italy*. SPRU Electronic Working Paper Series. 2007, Brighton: University of Sussex.
9. Martin, R. and P. Sunley, Complexity thinking and evolutionary economic geography. *Journal of Economic Geography*, 2007. **7**(5): p. 573–601.
10. Winterson, J., Alchemical city. *Radio 4*. 2014.
11. Axon, W.E.A., *The annals of Manchester: a chronological record from the earliest times to the end of 1885*. 1886, J. Heywood, Deansgate and Ridgefield.
12. Wyke, T., B. Robson, and M. Dodge, *Manchester: mapping the city*. 2018, Edinburgh: Birlinn Ltd.
13. Swinney, P. and E. Thomas, *A century of cities: urban economic change since 1911*. 2015, London: Centre for Cities.
14. Young, R.J. and P.A. Lovell, *Introduction to polymers*. 2011, Boca Raton: CRC Press.
15. Harris, P., *Salford at work: people and industries through the years*. 2018, Stroud: Amberley Publishing.

78 Rebuilding Urban Complexity

16. The Manchester Guardian, The chemical industry. *The Manchester Guardian*. 2 October 1926, Manchester.
17. Standring, P., Chemical industry's rapid progress in service of textiles. *The Manchester Guardian*. 1953, Manchester.
18. Riello, G. *The rise of calico printing in Europe and the influence of Asia in the seventeenth and eighteenth centuries*. 8th Global Economic History Network Conference on Cotton. 2005, Pune, India.
19. Sykas, P., *Pathways in the nineteenth-century British textile industry [Kindle edition]*. 2022, Abingdon, Oxon: Routledge.
20. Parsons, M. and M.B. Rose, The neglected legacy of Lancashire cotton: industrial clusters and the U.K. outdoor trade 1960–1990. *Enterprise and Society*, 2005. **6**: p. 682–709.
21. Tippett, L.H.C., et al., *The story of Shirley: a history of the Shirley Institute, Manchester, 1919–1988*. 1988, Manchester: The Shirley Institute.
22. Neffke, F., et al., Agents of structural change: the role of firms and entrepreneurs in regional diversification. *Papers in Evolutionary Economic Geography*, 2014. **14**(10).
23. Groom, B., *Made in Manchester: a people's history of the city that shaped the modern world*. 2024, Dublin: Harper Collins.
24. Clay, H. and K.R. Brady, *Manchester at work: a survey*. 1929, Manchester: The Manchester Civic Week Committee.
25. Wilson, J.F. and J. Singleton, The Manchester industrial district, 1750–1939: clustering, networking and performance. In *Industrial clusters and regional business networks in England*, J.F. Wilson and A. Popp, Editors. 2017, Abingdon and New York: Routledge.
26. Mitchell, T.M., *What Manchester did yesterday (the history of the establishment of the gas industry in Manchester)*. 1987, Manchester: British Gas North Western.
27. Marshall, A., *Industry and trade*. 1919, London: Macmillan.
28. Henriques, D.Q., *The textile industry of Lancashire*. 1952, Manchester: Manchester University Press.
29. Halls, J. and A. Martino, Cloth, copyright, and cultural exchange: textile designs for export to Africa at the national archives of the UK. *Journal of Design History*, 2018. **31**(3): p. 236–254.
30. Jones, K. and P. Allen, Historical development of the world rubber industry. In *Developments in crop science, vol. 23. Natural rubber: biology, cultivation and technology*, M. Sethuraj and N. Matthew, Editors. 1992, Amsterdam: Elsevier. p. 1–25.
31. Pinarbasi, S., Manchester antislavery, 1792–1807. *Slavery & Abolition*, 2020. **41**(2): p. 349–376.
32. Bullen, E., *Manchester migration: a profile of Manchester's migration patterns*. 2015, Manchester: Manchester City Council.
33. Haslam, D., *Manchester, England: the story of the pop cult city*. 2000, London: Fourth Estate.
34. Spinoza, A., *Manchester Unspun: pop, property and power in the original modern city*. 2023, Manchester: Manchester University Press.
35. New Economy, *Deep dive 0.2 manufacturing*. 2016, Manchester: Manchester University Press.
36. Harman, R. and J. Minnis, *Sheffield: Pevsner architectural guide*. 2004, New Haven and London: Yale University Press.
37. Jones, M. and J. Jones, *Sheffield at work: people and industries through the years*. 2018, Stroud: Amberley Publishing.
38. Potter, A. and H.D. Watts, Revisiting Marshall's agglomeration economies: technological relatedness and the evolution of the Sheffield metals cluster. *Regional Studies*, 2014. **48**: p. 603–623.
39. Moffat, A. and G. Rosie, *Tyneside: a history of Newcastle and Gateshead from earliest times*. 2006, Edinburgh: Mainstream Publishing.

How local economies evolve and branch 79

40. The Manchester Guardian, How chemists have helped the dyestuff industry. *The Manchester Guardian.* 22 January, 1927.
41. Martin, R. and P. Sunley, On the notion of regional economic resilience: conceptualization and explanation. *Journal of Economic Geography*, 2015. **15**(1): p. 1–42.
42. González, C., Evolution of the concept of ecological integrity and its study through networks. *Ecological Modelling*, 2023. **476**: p. 110224.
43. Nelson, R.R., A perspective on the evolution of evolutionary economics. *Industrial and Corporate Change*, 2020. **29**: p. 1101–1118.
44. Psarra, S., *The Venice variations: tracing the architectural imagination.* 2018, London: UCL Press.
45. Martin, R. and P. Sunley, Path dependence and regional economic evolution. *Journal of Economic Geography*, 2006. **6**(4): p. 395–437.
46. Meech, S., Fabrications: using knitted artworks to challenge developers' narratives of regeneration and recognise Manchester's South Asian working class textiles businesses. *Textile,* 2022: p. 1–22.
47. Froud, J., et al., *Coming back? Capability and precarity in UK textiles and apparel.* 2017, Manchester: Alliance Manchester Business School & School of Materials.
48. Hillier, B. and J. Hanson, *The social logic of space.* 1984, Cambridge: Cambridge University Press.
49. Jacobs, J., *The death and life of great American cities [1993 edition].* 1961, New York: The Modern Library.
50. Metcalfe, J.S., *Evolutionary economics and creative destruction.* Vol. 1. 1998, London: Psychology Press.
51. Rodgers, H.B., Manchester revisited: a profile of urban change. In *The continuing conurbation: change and development in Greater Manchester*, H.P. White, Editor. 1980, Gower: Farnborough. p. 26–36.
52. Deleuze, G. and F. Guattari, *A thousand plateaus: capitalism and schizophrenia.* 1987. Minneapolis: University of Minnesota Press.
53. Haslam, D., *Manchester, England: the story of the pop cult city.* 2000, London: Fourth Estate.
54. DeLanda, M., *A thousand years of nonlinear history.* 1992, New York: Zone Books.
55. Malm, A., *Fossil capital: the rise of steam power and the roots of global warming.* 2016, London: Verso Books.

7 How spatial complexity evolves and supports branching economies

Barabási's assertion that in order to understand a network, we need to understand how it was formed, again rings true when it comes to *spatial* networks in cities. Given that street systems generally evolve over hundreds – in some cases thousands – of years, we need to dig particularly deep to understand 'the morphological trajectory' of a city [1]. In this chapter we will begin by focusing on the architectural and planning theory associated with urban spatial evolution and go on to examine the morphological trajectory of three different cities up until the mid-20th century (Manchester, Sheffield and Newcastle). We will then consider how this evolving spatial evolution would have supported urban *economic* branching and evolution during the same period of time.

Emergence and self-organisation in urban street systems

Most cities were never formally planned. They evolved, developing incrementally through processes of self-organisation – a myriad of microscale building decisions which slowly came together to produce an emergent order: a *liveable* urban form. An early proponent of the importance of evolution to urban form was Patrick Geddes, who wrote *Cities in Evolution* in 1915 [2]. Geddes was originally a biologist, and he drew on evolutionary ideas from theorists such as Darwin and Huxley in his thinking on urban planning and design. However, he preferred to see human evolution as being based on processes of cooperation as opposed to competition or 'the survival of the fittest' [3]. Geddes stressed that urban development should be seen as a process of gradual "unfolding" as opposed to something which needed to be imposed all at once in a master plan. Although he achieved some influence over his contemporaries, such as Patrick Abercrombie and Lewis Mumford, he nevertheless failed to stop a general flourishing of top-down planning during the early 20th century, not least by Ebenezer Howard in his design of the "garden city".

Later in the 1960s, Jane Jacobs and Christopher Alexander touched on the importance of evolution in their writing about urban form, particularly at the local scale. Alexander identified how complex urban forms could emerge from local rules in "unfolding increments" (quoted in [3, p. 10]). However, neither of these authors described the incremental emergence of street patterns in any detail. Bill Hillier would later explore this process, based on the villages that he knew well in

DOI: 10.4324/9781003349990-10

the Vaucluse region of the South of France. He recognised a common set of patterns and structures in the villages and realised that he could reproduce these structures with computer models that followed very basic rules – such that new buildings should have open space in front of them, and that these buildings should be joined directly onto existing buildings, but not at corners. A third important rule was to avoid disrupting longer "lines of sight" or movement routes where a choice was available [4]. Systems theorists identify such local rules as "neighbour-neighbour" signals, and point to these as a common mechanism through which local interactions lead to global forms [5].

As the villages in Hillier's computer simulations grew, they slowly formed 'beady ring' structures which offered choices in how people (or in this case computer agents) could move around the street system. In the actual Vaucluse villages, these rings sometimes broadened out to accommodate a nascent public space. Hillier's work demonstrated that the *liveable* configurations with which we are all familiar in villages, towns and cities can therefore easily be derived from simple building decisions, as long as basic local rules are met.[1] Importantly, Hillier did not think that such rules were embedded in people's cognition (as the structuralist Claude Levi-Strauss or linguist Noam Chomsky might have thought). Rather, people "live the rules", paying attention to the patterns that already exist in the built environment around them and reproducing them through a process of "description retrieval".

More recently, Kim Dovey from the Melbourne Design School has been similarly exploring the emergence of local-scale structure in *informal and unplanned* settlements – which, as he identifies, is one of the dominant building processes occurring around the world today [6]. Dovey describes how, as new shelters are built in such settlements, they often gradually form linear streets, and then start branching though urban morphologies such as "combs" and "fishbones", often with blocks being only two plots wide on their shortest side (which he calls the 'two plot rule'). Through such processes, grids start forming. He finds, as did Bill Hillier, that buildings and public spaces co-evolve – and gradually routes form which allow people to navigate the space, which for Dovey is a principal dimension of how such places become "liveable" and avoid becoming slums. Nevertheless, so-called pirate landlords often illegally take over areas of land and preorganise them into grids, sometimes with less respect for the "two plot" rule (consequently producing urban forms with less liveability).

Self-organised urbanism creates local conditions – but also more global city forms. Rapidly organised informal settlements often evolve in a way which is segregated from broader city street systems – meaning that these new settlements are often *locally* integrated but globally cut off from the rest of the city [7, 8]. However, when cities evolve more slowly and incrementally as centres of trade and commerce, local rules to not disrupt longer lines of sight create the *foreground network* of streets with which we are already familiar from previous chapters. Such foreground networks are, for Hillier, a common and predictable feature of productive and commercial cities. In contrast, background networks of streets are more likely to be influenced by local social mores and the cultural conventions

82 *Rebuilding Urban Complexity*

which are embodied in local "spatial cultures". Analysing cities in the Maghreb region of North Africa, the architect Besim Hakim and Peter Rowe, for example, noted that complex and lively cities had grown up in this region based on simple Islamic "no harm" rules associated with the Sunni-Maliki school of law [9]. Common conventions to respect privacy, to preserve your neighbours' access to rights of way, and to ensure the proper harnessing and drainage of water had resulted in a very successful form of self-organised urbanism. When disputes arose, local judges intervened, but 'no exact dimension is ever specified, and the solution had to be determined on a case by case basis' [9, p. 24]. Hakim and Rowe's article has become a key reference for planners and architects who would today prefer a more bottom-up process to planning, based on "design codes", as opposed to top-down master plans – something that we will return to in Chapter 11.

Mapping the morphological trajectories of post-industrial cities

What relevance do these processes of self-organising urbanism have to the development of the post-industrial cities which are the concern of this book? In analysing the historical development of cities such as Manchester, Sheffield and Newcastle, it is clear that bottom-up self-organising patterns have dominated for long periods of history, interspersed with the imposition of top-down master plans. In Manchester and Sheffield it took until the 1960s for such master plans to have a major impact, but Newcastle was to experience a much earlier reconfiguration of its central streets [10].

As with the study of economic evolution, tracing the origins of street network morphologies is an exercise in "ancestry analysis" or perhaps rather "network archaeology". It requires what planners in the city of Sheffield called being able to 'read the past in the present landscape' – where former schemes of 'legibility' have become 'fossilised'.[2] This analysis is not always straightforward – as David Haslem describes for Greater Manchester, '[C]ities more than two centuries old are like a collage in their design; bits glued on, added, covered over' [11, p. xviii]. Nevertheless, as cities evolve incrementally, parts of their structure remain, which Julienne Hanson called 'morphological permanences' [1, p. 186].

A trading interface

In Greater Manchester, once an early trading point, a key "morphological permanence" which has remained important is the network of foreground streets in the city centre. The area, close to the Pennine Hills, was rich in water and coal [12]. Two adjacent urban settlements, Salford and Manchester, began to emerge on either side of the River Irwell, combining to form a key convergence point for travel between the farther-afield towns of Liverpool, Chester and York. The streets (Chapel Street, Deansgate and Oldham Street respectively) that brought people in and out from these other towns were still the most important arteries in the city's street system in the 1850s (see Figure 7.1) and remain important today, despite being disrupted by local infrastructural change.

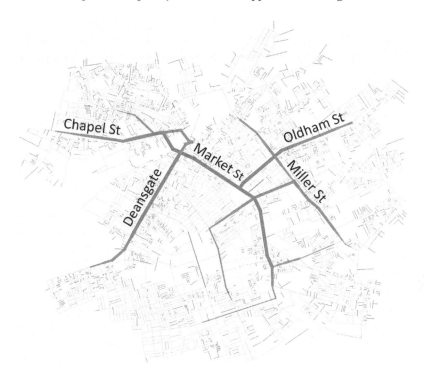

Figure 7.1 The foreground network of the 1850s city. This map highlights the top 5 per cent of streets in the 1850s city, according to a normalised version of the space syntax "choice" variable

Source: Author's own diagram, based on space syntax analysis of the Town Plan 1056 First Edition

In the 1850s, these foreground network "arteries" would have been particularly important in bringing strangers in from surrounding towns and regions, and then "slowing them down" [see 13, p. 17] along key central streets such as Market Street, supporting exchange in the city's central market squares. These streets would also have helped to promote the heightened information exchange so important to the industrial revolution – what Peter Hall called 'the first true innovative milieu' due to its 'capacity for continuous innovation through networking' [14, p. 307, 314]. As with Camillo Sitte's Italian medieval cities, the overall urban fabric of Manchester was largely constructed to support movement and trade at this time, with few public spaces except for those created in the widenings of central streets. Although the Town Hall was then relatively segregated in the urban fabric, the Manchester Royal Exchange was on the most integrated segment of the whole 1850s street network, with just such a widening of public space outside it where people could meet [15]. These busy central streets hosted not only many "third spaces" – pubs, inns and beer houses – but also churches, which became the focus of scientific and industrial discussion, helping to promote new forms of "technical literacy".

84   *Rebuilding Urban Complexity*

In Newcastle too, 'many layers and levels and heights and materials are woven into a rich fabric but Newcastle's medieval origins can still be found at its heart' (R. Fawcett. quoted in [10, p. 31]). The city grew up around the castle and the city's religious institutions, but trade was key, with an important foreground street being Bigg Market, which hosted wheat and meat markets as it widened on its way down towards the cathedral, the castle and the quayside. In the Georgian era (1820–30s) planner Richard Grainger and architect John Dobson joined forces to carve a new set of streets through what had previously been a relatively open space to the north of the city left by the demolition of a nunnery. This was a fairly unusual intervention at the time, as Georgian construction in other cities such as Bath had been more peripheral [10].[3] New elegant streets – including Grainger Street and Grey Street – were complemented by an indoor market which continues to operate today (see Chapter 11). John Dobson also designed Eldon Square at this time, a well-proportioned square which has personal meaning for me as my great-grandparents had a florist shop on the square in the early 20th century, where they sold produce from their nurseries to the north of the city. Overall, these interventions pushed the urban heart of the city away from the bustling quayside.

In Sheffield, it was the artisans who gradually moved the heart of the city away from the old civic and religious centre which had grown up around the Parish Church, Cutlers Hall and Paradise Square. The "topological centre" of the city shifted to newly constructed urban grids which were dominated by clusters of manufacturers and workers [16].

Grids: an infrastructure for complexity?

While the foreground network of cities often has a deep historical origin, the background network of residential streets often grows and changes more haphazardly. The medieval fabric of the combined urban area of Manchester and Salford grew into a deformed grid, for example, made more complex by backstreets, alleys and "courts", or courtyards. As this urban area expanded, more uniform grid systems appeared in some places as landowners rapidly converted empty plots into investment opportunities during the heady days of the industrial revolution. In 1775, the area of Ancoats, for example, was divided up in a grid pattern to provide units of land attractive in price to small builders [17] (see Figure 7.2). To the other side of Great Ancoats Street, the area now known as the Northern Quarter was also originally designed as a grid-based residential master plan, but it succumbed to a succession of commercial uses, from textiles weavers and artisans to large warehouses and then to the commercial activities present in the area today.

In Sheffield, the early medieval urban grid hosted long, narrow building plots known as "burbage plots", which allowed as many businesses as possible a commercial frontage onto the street [18]. Again, more uniform grids were later introduced by landowners such as the Duke of Norfolk, who sought to subdivide their land for sale using a system of leases – with these grids, as has already been

Figure 7.2 18th-century grid of streets in Ancoats
Source: Laurence, 1793 courtesy of the Digital Archives Association

described, hosting many different types of artisan right in the heart of the city. As in Manchester, areas which had once been planned as purely residential grids (such as the former deer park in the southeast of the city), again later diversified to accommodate factories, cutlery and tool-making workshops, more affordable back-to-back housing and alleyways and, later, larger-scale industry.

The fact that urban grids came to host such a diversity of land uses in both Manchester and Sheffield is perhaps no coincidence. A number of theorists have noted that the oft-maligned grid is actually a very useful structure for the incremental generation of lively urbanity. The architect Leslie Martin, for example, suggested that grids 'afford complexity', acting like a net which is thrown on the ground, which over time supports the development of more intricate spatial configurations. In his celebrated article 'The Grid as Generator' [19], Martin describes the grids that were laid out as part of the rapid development of cities in the United States – which, while appearing uniform, create anything but uniform outcomes when it comes to urbanism. The architect Arnis Siksna [20] compared these grids with those found in Australian urban centres, again noting how quickly they supported architectural complexity as new urban blocks were added *within* existing blocks. This evolutionary process is also clearly visible in the historical development of the post-industrial cities surveyed both here and in the United States. As Sheffield and Newcastle grew, for example, original grid-like burbage plots were gradually intensified and infilled.[4] Rae describes how in New Haven, the existing grid went from nine to twenty-nine squares between 1784 and 1802 as each original

86   *Rebuilding Urban Complexity*

square was bisected by a new street. This resulted in a set of 'nested squares' which have endured through the centuries, acting as 'perceptual anchors' in the city [21, p. 38]. As the central grid of the city intensified, it also increased its active commercial frontage by some four kilometres.

Bill Hillier was keen to point out that grids still structure movement at *a more global scale*, in part because they are constituted by both shorter and longer lines. Manhattan, for example, is made up of a set of intersecting (shorter) *streets* and (longer) *avenues*, with the avenues providing connections to the more global city, meaning that they are more likely to be "foreground network" streets. Hillier also pointed out that certain lines in grids are better connected to the "outside" of the city [22]. Nevertheless, he felt that the more deformed a grid was, the more intelligible it would be to people, with Haken and Portugali also suggesting that 'broken symmetries' allow cities to 'afford remembering' [23]. The (deformed) grids of urban street systems can be seen to provide a key infrastructure for evolving complexity.

The intermingling of industries at the local scale

Within these deformed grids, however, other morphological "affordances" are important to how economic diversity becomes mixed and arranged at a very fine grain. A developing line of space syntax research (associated with authors such as Alan Penn, Laura Vaughan and Sam Griffiths) has focused on how the spatial configuration of cities supports patterns of *interrelated* urban diversity – hosting not a random mixing, but rather a closely interdependent set of economic activities [24, 25]. It is pointed out that a key element underlying socio-economic diversity is 'morphological diversity', which is important, for example, to how high streets in suburban areas both produce land use diversity and then remain resilient over time [25].

A glance at historic maps reveals that the griddy background-network streets of Manchester, Sheffield and Newcastle hosted a teeming economic diversity of land uses from well before the Victorian period. Supporting *morphological* diversity is also visible across the urban fabric. An area of what is now known as the Northern Quarter in Manchester is captured, for example, at a certain part of its history by the GOAD fire insurance maps, which were drawn up to assess fire risk and hence are meticulous in their detail. The plan for this part of Manchester reveals a fine-grained co-location of economic activities – from hat, umbrella and cushion factories to shops and warehouses which are all mixed in with dwellings (see Figure 7.3). Archival research in trade directories adds important underlying detail to the map – revealing, for example, that Drey, Simpson & Co Ltd, which managed the home trade warehouse (H.T.W) on Newton Street, were a Stockport-based firm that made specialised finishes for velvets.[5] Archibald Winterbottom & Sons in the same block were dyers and finishers, while also manufacturing bookbinding cloth[6] – the locations of these two firms are marked on the map. The GOAD maps also reveal the different affordances offered to industry at the local scale – these

How spatial complexity evolves and supports branching economies 87

Figure 7.3 Land uses in the area around Dale Street, 1886: annotated GOAD Fire Insurance Plan

Source: This map was produced by Nicolas Palominos using as a base map the Fire Insurance Plan of the City of Manchester, published by Chas E Goad, 1886, Volume 1, sheet 7. British Library Collection in the public domain via Wikimedia Commons

included cranes, wharves, yards, weighing machines and multiple storage spaces for raw and waste materials (something that we will return to in Chapter 10). The area also offered its workers dining rooms, public houses and a hospital.

An additional shared "affordance" for local businesses in this part of Manchester was, of course, the locally interconnected street system – once a residential grid which had become overtaken by commerce and industry. We have already noted the back streets that assisted deliveries and provided cheaper spaces for production and storage; Back China Lane is parallel (to the east) of China Lane as indicated on the second map of this area (Figure 7.4). The canal would also have provided a conduit for goods and materials, while Dale Street, as a foreground network street, would have brought people into the area to trade. The area also boasts a variety of plot sizes (Figure 7.4) – a factor which a number of authors identify as supporting local economic complexity – not only because different sizes of plot support different industries, but also because this provides a variety of commercial spaces which economic activities can move into as they expand [25, 26].

Figure 7.4 Building footprint in the area around Dale Street, 1886

Source: This map was produced by Nicolas Palominos using as a base map the Fire Insurance Plan of the City of Manchester, published by Chas E Goad, 1886, Volume 1, sheet 7. British Library Collection in the public domain via Wikimedia Commons

The GOAD maps also reveal finer forms of spatial organisation, showing, for example, how local business used the different network properties of the street system to arrange themselves in ways that would minimise mutual disturbance. For example, in Figure 7.3, Dale Street can be seen to host a line of offices, while dwellings are found around the corner on China Lane. This "round the corner" linear positioning is also common today in how economic diversity organises itself at the very local scale [22] – something which is not always taken account of in "mixed use" housing developments.

While we cannot directly discover this from the GOAD map itself, the map is evocative of 'the emergent patterning of everyday routines' and the '"noise" of everyday urban life' which the morphological and economic diversity of this area would have supported [27, p. 490]. The different rhythms of overlapping social and economic practices (working, dwelling, dining) would have played a role in keeping this area lively at different times – something which Jane Jacobs held to be a particularly important aspect of successful urbanity. These rhythms

How spatial complexity evolves and supports branching economies   89

of intersecting life were also highlighted by the celebrated French urban theorist Henri Lefebvre, who talked about the need to do 'rhythmanalysis' in order to understand cities [28].

Archival research in other parts of Greater Manchester also reveals how the "differential accessibility" of the street system was exploited by firms in the management of their diverse *internal* production chains. In 1892, for example, the firm Kendal Milne comprised cabinet makers, upholsterers, general house furnishers and drapery warehousemen. This company, which began as a shopping bazaar in 1796, at this time hosted two large stores on Deansgate (a key foreground network street), one focused on cabinet furniture and upholstery and another on fashion and drapery (with the latter going on to become House of Fraser). Around the back of these buildings on Garden Lane a cabinet factory employed about a hundred workers in a seven-storey building which accommodated sawmills, a timber measuring room and cabinet-making workshops. A few turnings away on Back Bridge Street, an upholstery works employed about 30 skilled people, while nearby on Wood Street (appropriately) there was a timber yard and extensive stabling for the horses that facilitated delivery [29].[7]

Expansion

As the cities of Manchester, Sheffield, Newcastle and (across the Atlantic) New Haven expanded following their industrial success, they largely preserved their morphological "backbones" and the essentially griddy structure of their back-streets. All four cities expanded at speed. The city of Manchester, for example, saw rapid expansion between 1850 and 1950, with the landed gentry selling land for the construction of not only factories but also the railways [30]. The "steel city" of Sheffield expanded from 91,000 people to more than 400,000 between 1831 and 1901 [18, p. 5], while New Haven saw an eightfold growth in population between 1850 and 1920, coinciding with a 'high-crested wave of money' washing over the built environment [21, p. 68]. Much of this rapid development again took the form of grid-like terrace streets which later became synonymous with Manchester through the television programme *Coronation Street*.[8] The quality of construction was often compromised in the rush, and a severe lack of supporting infrastructure (sewage and water) led to unsanitary, and often overcrowded, living conditions, as we will later explore further.

Nevertheless, in Manchester, the griddy new urban areas of the city knitted well into the distributed and open network density already present in the city centre. As the city grew, its foreground network also extended out into its growing urban fabric, linking the new streets into an overall coherent structure, and becoming itself the focus of very lively commercial activities. The industrial commentator Shadwell remarked at the turn of the 20th century that 'in the main arteries where the tide of life runs at the full, it runs with a roar and a stir and a bustle which are not excelled by any other town, not even by New York or London itself' [31, p. 66].

90   *Rebuilding Urban Complexity*

Exploring the expansion of Sheffield between 1770 and 1905, Sam Griffiths also found that the intelligibility of the city was sustained across all scales, in part through the embedding of new streets into the previously existing configuration of the rural hinterland. A global "centre-edge structure" was maintained, while more localized centres were also preserved to ensure what Hillier would have called "pervasive centrality" – importantly, 'each "phase" of urbanisation was premised upon structural continuities bequeathed by previous phases' [32, p. 7].[9]

Manchester's foreground network had by this time been reinforced by a layering of other transport networks – from the canal system which bought industrial raw materials right into the heart of the city, to the railway system which for the first time was beginning to link the UK's cities into a coherent grid, to the Ship Canal which connected the city into global trade networks. At the more local scale, the city also had a tram system, which closely followed the foreground network and extended the "integration core" of the city, providing people with a quick means of accessing all parts of the urban fabric (the old tram system had a much wider coverage of the city than the new Metrolink which is nevertheless a popular complement to the street system today). The cities of Sheffield and Newcastle equally benefited from tram and trolley bus systems which provided relatively affordable but rapid forms of public transport (a form of street-based transportation which many cities now sorely miss). In his documentation of the rise to prosperity of New Haven, Douglas Rae also notes the power of this 'layering' of different forms of 'network centrality' associated with transport infrastructure. He describes how completion of the railroads in the United States connected the cities of New Haven, Detroit, Chicago, New York and Philadelphia into a closely woven network of railroad hubs which would 'bind the rest of the nation into its orbit' (D.W. Meinig quoted in [21, p. 66]). More locally, Rae describes how the railroad lines formed a radial geometry around the "old nine squares" of the street grid, creating 'pulsing arteries of industrial supply and distribution' [21, p. 68]. In Manchester, the canals, railways and streets also interwove into a dense fabric of supply, production and distribution – particularly in industrial areas such as Ancoats.

How evolving spatial systems support *economic* branching and evolution

As these industrial cities built additional layers of infrastructure, the close juxtaposition of economic sectors at the local scale would have created a situation ripe for the cross-fertilisation of ideas and the path interdependent economic branching described in Chapter 6. The overall distribution of industrial diversity remained remarkably stable in Manchester from the Victorian period onwards, being replicated across the urban fabric, providing an affordance for the local cross-sector fertilisation of ideas and creativity right across the city. The local intermingling of industries in Manchester persisted well into the 1950s – see, for example, the map in Figure 7.5, which shows an area partially enclosed by a loop in the River Irwell which hosted manufacturers of carpets, wire, gears, rubber, cotton waste and waterproof clothing, amongst others.

How spatial complexity evolves and supports branching economies 91

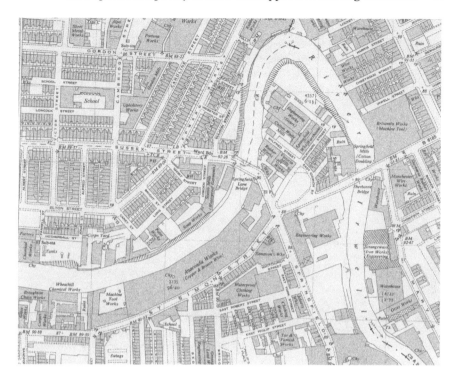

Figure 7.5 Diverse land uses south of the River Irwell in Broughton, 1952

Source: Ordnance Survey 1:2500 scale map, 1952, Creative Commons Attribution (CC-BY) license. Reproduced with the permission of the National Library of Scotland (see https://maps.nls.uk/. Accessed [12/09/24])

The generic properties of space

As the economies of these different cities evolved, their overall spatial configuration and morphology proved adaptable to successive waves of economic change – from artisanal production to steam-powered mills and then electrically-powered factory systems and from commercial warehouses to retail-focused department stores. Bill Hillier suggested that while economic functions change rapidly, urban spatial form changes slowly, with the generic networked properties of space proving adaptable across multiple types of urban economy. The street system does not need to be reinvented each time a new economic "paradigm" comes along. While individual industries might have changed, the configurational spatial properties of each of the cities reviewed here remained very similar until the 1950s – with their fine-grained juxtaposition of economic diversity at the local scale, supported by gridlike street systems, joined up through citywide foreground networks. Douglas Rae describes, for example, how 'New Haven would deposit its industrial capital – made visible in brick and steel and audible as clang and clatter – over and around a quieter grid created before an industrial city could even have been envisioned'. The adaptability

92 *Rebuilding Urban Complexity*

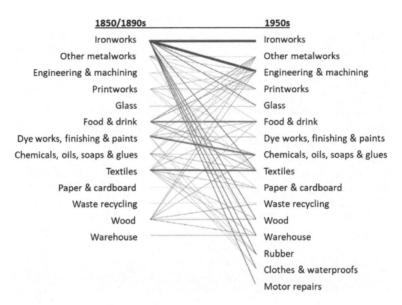

Figure 7.6 Change in use of plots between 1850/1890 and 1950. The thickness of the line demonstrates how many times the 1850/90s use transformed into the 1950s use – the thickest line being 8 times and over, the medium line 5–8 times and the thinnest line 1–4 times

Source: Author's own diagram

of the underlying grid meant that 'New Haven's central district has organized the city's mental map of itself for 360 years' [21, p. 37].

Adaptability to new economic evolution at the city scale was also replicated at the local building plot scale. Individual plots were used and reused by changing industries during the evolution of Greater Manchester, for example. Former ironworks were used for a particularly diverse set of industries (see Figure 7.6). Both buildings and plots have thus proved *pre-adaptive* to multiple types of economic use.

The adaptability of streets (and their surrounding backstreets) to successive economic change is also richly illustrated by an example in London – the "furniture street" of Tottenham Court Road (see Box 6).

Box 6 Tottenham Court Road – home to furniture since the 18th century

Tottenham Court Road and its surrounding streets have been a "furniture cluster" in London since the late 18th century. However, in this time, the area has supported many different modes of production and distribution – from small artisans to huge integrated stores and factories, to now mostly retail stores. In fact, it is not the road which has been important in isolation, but a spatial niche incorporating both the road and its backstreets, with the road being exploited for its footfall and the backstreets being used for production.

The heyday of this economic niche was between 1850 and 1950 [33]. In the early 19th century, the road was on the periphery of London, which made production space relatively cheap. When the predecessor of the furniture store Heals first came to Tottenham Court Road in 1818, the lease stipulated that there should be appropriate accommodation for 40 cows.[10] From the mid-19th century there were 30–40 furniture businesses on Tottenham Court Road, with many more coming to occupy in the backstreets, with particular local sub-clusters such as cabinet makers around Charlotte Street [33]. By the 1880s these smaller artisans and shops had consolidated into some much larger concerns, including Maples, which employed 2,000 staff, with 1,295 in factories onsite and in backstreets such as Beaumont Place, Midford Place, Frederick Street and Southampton Court. The buildings in these streets still have the hooks on them for the pulleys that were used to bring goods to the upper floors. The Maples buildings on Tottenham Court Road itself (at the corner with Euston Road) gradually took over an area once occupied by 200 homes.

Helped by the arrival of Goodge Street and Warren Street tube stations in 1907, the street supported more and more retail as it became central to the broader city. Furniture is an example of a product which has a "high threshold" population (being an expensive item that people do not buy regularly), and a "high range" (as people are prepared to travel far to buy it) so furniture merchants often end up centrally located and clustered in cities, offering their customers a diversity of choice [24, 34]. The retail cluster remains today, despite the competition offered by large out-of-town stores such as Ikea (and indeed Ikea for a while itself opened a small showroom on the road). Around the corner from Heals, some of the old industrial buildings now host creatives from the local TV, media and advertising cluster, which has long been based to the north of Soho. Indeed, these old production spaces are also proving useful for contemporary creative production, with their high ceilings, good light, and "edgy" warehouse branding. Further north along the road, Mitford Place, which once housed production for Maples, now hosts a pharmaceutical-orientated artificial intelligence company not far from a cluster of pharmaceutical companies which have formed around neighbouring University College Hospital. It is not just the furniture sector niche which has survived the test of time, but also this retail-production, foreground-background street arrangement which so effectively exploits the spatial affordances of the local street system; the interconnectedness of this niche within the wider local urban fabric; and the centrality of this location within London as a whole.

94 *Rebuilding Urban Complexity*

The importance of "messy" redundancy and inefficiency

These examples reveal how the basic set of "constraints" which allow liveable street systems to emerge from processes of self-organisation also clearly make these same street systems adaptable to economic evolution and change. Another important way in which the built environment supports economic change is through the existence of a degree of *redundancy* in the system – returning to a characteristic of complex adaptive systems which was discussed in Chapter 2.

As an example of such redundancy, Jane Jacobs underlined the importance of *older* buildings in cities as resources for the experimentation and innovation that is essential to the economic branching process. She famously argued that 'old ideas can sometimes use new buildings. New ideas must use old buildings' [35, p. 245]. This is in part because older buildings are often more affordable for new firms. However, older buildings also provide the physical conditions in which material problem solving and experimentation can take place without the fear of damaging expensively designed interiors. They thus form part of what Jacobs [36, p. 85] describes as the 'valuable inefficiencies and impracticalities of cities'.

In Manchester, the railway arches associated with the elevated viaducts of the Victorian railways produced residual spaces which are now proving useful and adaptive to the needs of manufacturing firms seeking cheaper and "messier" space in the heart of the city, in a similar way to elsewhere in the UK (see Box 2) [37]. The textile mills which remained standing after the period of manufacturing decline in the mid-20th century have also provided valuable "left-over" spaces for new forms of experimentation. Because of their size, the mills could support multiple small businesses, offering a useful site of small business incubation and the potential for knowledge spillovers and shared support systems [38]. One group of entrepreneurs who took advantage of these spaces from the 1960s and 1970s onwards was a Pakistani network of knitwear producers, who also appropriated the machines and technologies which had been abandoned along with the mills.[11] They supplied well-known British brands such as Marks & Spencer, Debenhams and British Home Stores, and their clustering in buildings in Ancoats, Ardwick and Cheetham Hill ensured ongoing synergies between the firms. One building in Chapeltown Street once hosted over 30 knitwear factories, while 40 knitwear factories were housed in two buildings in Dolphin Street [15]. Because of this clustering, buyers were able to come and see several firms at the same time, and a "social economy" developed, based not on cash but on favours. The companies also shared technicians, with one technician, for example, working at one factory until 12 pm and then at another from 12 to 6 pm. On Friday the company workers were able to take advantage of a nearby mosque for prayers [39]. While the knitwear sector in Manchester has somewhat fallen under the radar in terms of its value to the local economy, this is an example of how older, less expensive commercial buildings can preserve what Peter Sunley et al call 'relic paths' [40, p. 388] which provide an additional interface between old and new capabilities, creating sometimes unlikely 'circuits of innovation' [14]. In the next chapter we will explore how

How spatial complexity evolves and supports branching economies 95

a combination of older buildings and technologies supported new branching out of the waterproofing industry – an industry which successively reinvented itself and still remains competitive today.

In summary

This chapter has revealed the gradual processes of self-organisation which are important to bottom-up, incremental city building. It has described how the underlying urban morphologies of English industrial cities developed largely bottom up until the mid-20th century, excluding the more ambitious city-centre master planning associated with Georgian interventions in Newcastle. During their industrial heyday, these cities brought traders together in their foreground networks and supported a tight intermingling of economic diversity in their griddy backstreets. This local intermingling supported the cross-sector exchange of capabilities so important to the heyday of these industrial cities, spanning the urban fabric. The *generic* nature of urban configurations (i.e., their interlocking foreground and background networks) also helped cities to adapt to economic change, while inefficiencies and redundancies in street networks and buildings have supported both the preservation of old economic paths and the experimentation so important for new ones. In the next chapter we will explore these themes of economic branching, path interdependency and spatial evolution in finer detail through a more specific case study from Manchester: the waterproofing industry.

Notes

1 Bill Hillier used to demonstrate this to his students at University College London using sweets, asking them to arrange the sweets only according to local rules (e.g. in terms of their relationship to their neighbouring sweets), and to observe how this gradually led to the generation of urban forms, without the need for an overarching design or master plan.
2 http://sytimescapes.org.uk/zones/sheffield/S12. Accessed [07/06/2024].
3 Although the elegant St Ann's Square in Manchester was also a product of the Georgian era.
4 http://sytimescapes.org.uk/zones/sheffield/S12. Accessed [07/06/2024].
5 Grace's Guide to British Industrial History. www.gracesguide.co.uk/. Accessed [07/06/2024].
6 As listed in the 1891 *Directory of Cotton Mills* in Manchester and Salford.
7 www.housefraserarchive.ac.uk. Accessed [07/06/2024].
8 See Brian Groom's 2024 book *Made in Manchester* for more precise details about the type of street construction which became famous through *Coronation Street*, a series created by Granada Television and shown on ITV since 1960.
9 Such maintenance and reinforcement of global spatial structure was also found by Sophia Psarra in her fascinating account of the expansion of the mercantile city of Venice, *The Venice Variations* (2018).
10 www.heals.com/blog/heals-200-tottenham-court-road. Accessed [07/06/2024].
11 With thanks to Numan Azmi, founder of the Manchester Knitters' Association, for providing information about this network of firms and its history.

96 *Rebuilding Urban Complexity*

References

1. Hanson, J., *Order and structure in urban space: a morphological history of the City of London*. Bartlett School for Architecture and Planning. 1989, London: University of London.
2. Geddes, P., *Cities in evolution: an introduction to the town planning movement and to the study of civics*. 1915, London: Williams.
3. Batty, M. and S. Marshall, Thinking organic, acting civic: the paradox of planning for cities in evolution. *Landscape and Urban Planning*, 2017. **166**: p. 4–14.
4. Hillier, B., The genetic code for cities: is it simpler than we think? In *Complexity theories of cities have come of age*, J. Portugali, et al., Editors. 2012, London: Springer. p. 67–89.
5. Wohl, S., From form to process: re-conceptualizing Lynch in light of complexity theory. *Urban Design International*, 2017. **22**: p. 303–317.
6. Dovey, K., et al., *Atlas of informal settlement: understanding self-organized urban design*. 2023. London: Bloomsbury Publishing.
7. Parham, E. The segregated classes: spatial and social relationships in slums. 8th International Space Syntax Symposium. 2012, Santiago.
8. Karimi, K., et al., *Evidence-based spatial intervention for regeneration of informal settlements*. 6th International Space Syntax Symposium. 2007, Istanbul.
9. Hakim, B.S. and P.G. Rowe, The representation of values in traditional and contemporary Islamic cities. *Journal of Architectural Education*, 1983. **36**(4): p. 22–28.
10. Barke, M., B. Robson, and A. Champion, *Newcastle upon Tyne: mapping the city*. 2021, Edinburgh: Birlinn.
11. Haslam, D., *Manchester, England: the story of the pop cult city*. 2000, London: Fourth Estate.
12. Groom, B., *Made in Manchester: a people's history of the city that shaped the modern world*. 2024, Dublin: Harper Collins.
13. Hillier, B. and J. Hanson, *The social logic of space*. 1984, Cambridge: Cambridge University Press.
14. Hall, P.G., *Cities in civilization: culture, innovation, and urban order*. 1998, London: Weidenfeld & Nicolson.
15. Froy, F., *'A marvellous order': how spatial and economic configurations interact to produce agglomeration economies in Greater Manchester*. Bartlett School of Architecture. 2021, London: University of London.
16. Griffiths, S. and K. Navickas, The micro-geography of political meeting places in Manchester and Sheffield c. 1780–1850. In *Micro-geographies of the Western City, c. 1750–1900*. 2020, London: Routledge. p. 181–202.
17. Rose, M.E., K. Falconer, and J. Holder, *Ancoats: cradle of industrialisation*. 2015, Swindon: Historic England.
18. Jones, M.J.J., *Sheffield at work: people and industries through the years*. 2018, Stroud: Amberley Publishing.
19. Martin, L., The grid as generator. In *Urban space and structures*, L. Martin and L. March, Editors. 1972, Cambridge: Cambridge University Press. p. 6–27.
20. Siksna, A., City centre blocks and their evolution a comparative study of eight American and Australian CBDs. *Journal of Urban Design*, 1998. **3**(3): p. 253–283.
21. Rae, D.W., *City: urbanism and its end*. 2005, New Haven: Yale University Press.
22. Hillier, B., *Space is the machine: a configurational theory of architecture*. Vol. 2007 e-print. 1999, Cambridge: Cambridge University Press.
23. Haken, H. and J. Portugali, The face of the city is its information. *Journal of Environmental Psychology*, 2003. **23**: p. 385–408.
24. Penn, A., I. Perdikogianni, and C. Motram, Generation of diversity. In *Designing sustainable cities*, G.E.R. Cooper and C. Boyko, Editors. 2009, Chichester: Wiley Blackwell.

25. Vaughan, L., et al., *An ecology of the suburban hedgerow, or: how high streets foster diversity over time*. 10th International Space Syntax Symposium. 2015, London.
26. Marcus, L., Spatial capital: a proposal for an extension of space syntax into a more general urban morphology. *The Journal of Space Syntax*, 2010. **1**(1): p. 30–40.
27. Griffiths, S. and L. Vaughan, Mapping spatial cultures: contributions of space syntax to research in the urban history of the nineteenth-century city. *Urban History*, 2020. **47**: p. 488–511.
28. Lefebvre, H., *Rhythmanalysis: space, time and everyday life*. 2004, London: A&C Black.
29. Richardson, N., *Good value and no humbug: a discourse on some of the principal trades and manufactories of Manchester*. [1892] 1981, Manchester: Self Published.
30. O'Reilly, C., Re-ordering the landscape: landed elites and the new urban aristocracy in Manchester. *Urban History Review/Revue d'histoire urbaine*, 2011. **40**(1): p. 30–40.
31. Shadwell, A., *Industrial efficiency: a comparative study of industrial life in England, Germany and America*. 1906, London: Longmans, Green, & Co.
32. Griffiths, S., *Persistence and change in the spatio-temporal description of Sheffield parish c.1750–1905*. 7th International Space Syntax Symposium. 2009, Stockholm.
33. Edwards, C., Tottenham court road: the changing fortunes of London's furniture street 1850–1950. *The London Journal*, 2011. **36**(2): p. 140–160.
34. Christaller, W., *Central places in Southern Germany [1933 edition]*. 1966, Englewood Cliffs, NJ: Prentice-Hall.
35. Jacobs, J., *The death and life of great American cities [1993 edition]*. 1961, New York: The Modern Library.
36. Jacobs, J., *The economy of cities*. 1970, New York: Vintage Books.
37. Froy, F. and H. Davis, Pragmatic urbanism: London's railway arches and small-scale enterprise. *European Planning Studies*, 2017. **25**(11): p. 2076–2096.
38. Lloyd, P.E. and C.M. Mason, Manufacturing industry in the inner city: a case study of Greater Manchester. *The Royal Geographical Society*, 1978. **3**(1): p. 66–90.
39. Meech, S., Fabrications: using knitted artworks to challenge developers' narratives of regeneration and recognise Manchester's South Asian working class textiles businesses. *Textile*, 2022: p. 1–22.
40. Sunley, P., R. Martin, and P. Tyler, Cities in transition: problems, processes and policies. *Cambridge Journal of Regions, Economy and Society*, 2017. **10**(3): p. 383–390.

8 Waterproofing

A case study

As has been discussed throughout this book, economic sectors that coexist in the same city benefit from cross-sector synergies as they grow and adapt. This reveals a form of place-based *path interdependency* [1]. A good example of such path interdependency is the waterproofing sector in Manchester, which evolved through a symbiotic interaction between the textiles, clothing, gas and chemicals industries – all sectors which were revealed in Chapter 3 to exhibit "industry relatedness" today. The waterproofing industry has evolved through many twists and turns, taking advantage of different properties of the built environment, including the strong local spatial cultures which existed in neighbourhoods in the north of the city. In this chapter we will focus on this case in more detail, revealing that ultimately this path interdependency expanded spatially to also incorporate the neighbouring city of Sheffield.

The birth of the macintosh

In fact, this is a tale of three cities, as the inventor of the first raincoat or "mac" was not initially based in Manchester but in Glasgow, which was then a centre of industrial chemistry. Charles Macintosh discovered how to make waterproof fabric in 1823 while experimenting with a waste product from gas (naphtha or coal tar) for his father's dye works – engaging in the widespread Victorian practice of reusing waste materials, which we will return to in Chapter 10.

Macintosh realised that he could use a naphtha-based rubber solution to produce a waterproof layer between two layers of fabric [2]. He moved to Manchester to fully capitalise on his invention, starting to work with cotton spinners at the Birley Mills in Chorlton-upon-Medlock to produce waterproof rainwear. Manchester had already built up the necessary capabilities for Macintosh's industry to flourish – not just in terms of the skills of workers, but also in terms of the embedded information and knowledge in buildings, factories, warehouses and specialised forms of capital investment.

The path towards wearable and practical waterproof clothing was a "problem-solving" one. A breakthrough came when the engineer and coach-maker Thomas Hancock joined Macintosh and improved his production process through "vulcanisation" [3].[1] Hancock had been inspired to make waterproof clothing

DOI: 10.4324/9781003349990-11

Waterproofing: a case study 99

when seeking new ways to protect passengers as part of his family's coach-making business. He initially chose not to patent his vulcanisation machine, giving it the unlikely name of "pickle" to disguise its functionality [4]. Nevertheless the underlying manufacturing processes spread across the city, creating what the economist Joseph Schumpeter would call a "swarm" of new waterproof companies [5]. Writing in the journal *Textile History*, Sarah Levitt provides a detailed account of the history of the rubberised garment trade in Manchester and Salford [6], revealing that by the 1890s there were seventy local macintosh factories, while by the early 20th century the city was responsible for two-thirds of the UK production of waterproof garments. The city was soon also exporting internationally, with a Chicago-based clothing catalogue advertisement urging, 'Get into the swim, Macintoshes are all the rage'.

The waterproofing industry continued to problem-solve and branch into new lines of work, with Levitt identifying progressive improvements in lightness, odour, colours, wearability and affordability. New firms developed to exploit the multiple different potentials associated with the functionality and style of waterproof fabrics. Both Jane Jacobs and Stuart Kauffman would have appreciated how new innovations in the industry opened the way for complementary additional innovations, through a continuous branching process – with each new product creating a foothold for still something else. One particular challenge which needed to be overcome was the poor breathability of early waterproof fabrics, which created a whole range of innovations, leading eventually to advanced materials such as Aertex. However, firms also experimented with the weave and weight of fabric, types of sewing silk, different linings, and varieties of colour and texture. Interestingly, each new "diversification surface" was very rooted in the material world – not only in textile materials themselves but also in the affordances of human bodies and the surrounding physical environment (see Figure 8.1).

When it came to the affordances of the human body, there were detachable collars and hoods, and new seams, joins and ventilating tubes to help with breathability. Adaptations to the physical environment included different fabrics for different types of precipitation – with Manchester producers specialising in *rain*proof garments, while down in London, the focus was on *shower*proof coats (developed by well-known brands such as Burberry) [6]. The waterproofing industry also found new directions during the First and Second World War. As an example, Perseverance Mills produced barrage balloons during the First World War before going on to develop computer ribbons and finally the waterproof Pertex fabric [3]. During the Second World War, J. Mandelberg & Co. built on their waterproof-based capabilities to produce flying suits and gasproof clothes, before later branching into hospital sheets and then hot water bottles [7]. Both examples illustrate how branching processes happen as much *inside firms* as they do outside of them – with firms often branching into "skills-related" products.

Another important waterproof fabric that was developed during the Second World War was Ventile – a fabric which is of interest as a more sustainable waterproof fabric today. The city's Shirley Institute was asked by Winston Churchill to develop a material that would help pilots to survive in freezing seas if their planes

100 *Rebuilding Urban Complexity*

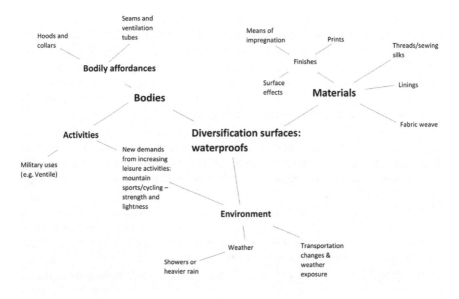

Figure 8.1 Diversification surfaces in the history of waterproofs
Source: Author's own diagram

went down. The result – Ventile – is pure cotton, but woven specially so that the fibres swell when wet, becoming impregnable in contact with water. Ventile is still used by waterproof coat makers in the city such as Private White V.C. (see Box 7), with the simplicity of the fabric offering greater recyclability than waterproof products based on more complex coatings and finishes.

Box 7   Private White V.C. – a brave manufacturer?

Private White V.C. is a luxury outerwear company which still operates from mill buildings in Strangeways. The company focuses on "slow fashion" – offering mainly hand-made products to a predominantly international market, spanning 30 countries. It has recently brought many distribution and retail processes in-house, preferring to sell directly through its own stores (in the factory and in Mayfair, London) and online. The company is nevertheless relatively unique in that it prides itself on working with local suppliers, with 80 per cent of its supply chain being found within sixty-four kilometres. Mike Stoll, one of the managers, identifies that there are thirty mills around them, particularly to the northwest of the city, and says the firm deal with seven or eight of them, including Mallalieus in Delph (in Oldham) and Marling and Evans in Slaithwaite. Other suppliers include local finishers and dyers – there are finishers, for example, who do showerproofing, stainproofing and waterproofing in Oldham, and in the region beyond Huddersfield. The embeddedness of the company in local supply chains has brought benefits. Mike has built his network over the past fifty years, and in the process has made many friends and absorbed a lot of knowledge. Their suppliers influence what they make – coming to them with product innovations and requests to develop things for them in the factory.

The company is based in a three-storey historic mill building with which the current owners, partners James Eden and Mike, have a long-term connection. James' great-grandfather, Jack, served his apprenticeship there, before serving as a private in World War 1 and being awarded the Victoria Cross for bravery. Mike has set up multiple brands in the building, including Cooper & Stollbrand. Both the buildings and these individuals therefore represent a continuity of waterproofing capabilities and knowledge in this part of the city – despite the slow disappearance of these capabilities all around them, with few manufacturers sharing their riverside location today.

Source: The information in this Box is based on an interview carried out during my PhD research – see [8].

As it developed, the waterproofing industry maintained a 'sustained dialogue across sectoral boundaries' [3, p. 706]. Much of this dialogue was based on social ties and friendships – something that we will revisit in Chapter 10 – with one manufacturer explaining, '[T]wo of my best friends are fabric importers and they brought in the fabric, and another of my very close friends has a coating plant' [3, p. 701]. Such social and family ties have also been important to business succession and the persistence of waterproofing capabilities in the city over generations – with Mike Stoll of Private White V.C. noting that he inherited the business from his father's best friend, with the business now going to his best friend's son [9].

The spatial organisation of waterproofing

The waterproofing industry clearly demonstrates the evolutionary interdependent branching processes which were so important to industrial development in both Greater Manchester and other industrial cities. What were the historic spatial foundations for this branching process? The mills where Charles Macintosh based himself on arrival in Manchester were in the heart of the city, just south

102 *Rebuilding Urban Complexity*

of the River Medlock, and right next to a gas works (although it was not initially possible to exploit the waste product of this gas works – naphtha – due to local regulations) [10]. As the waterproofing industry took off, it spread across the urban fabric and started to cluster to the north of the city, in the Cheetham Hill and Broughton area of Salford (which hosts the Strangeways area described in Chapter 5). By the 1930s, some two-thirds of Salford's new clothing plants produced waterproof clothing [11]. The new waterproof companies sometimes used old buildings, but analysis of historical maps suggests that they also developed new premises in green field sites or previously residential sites, including "fitting into the gaps" between residential buildings. The rubber factories, in comparison, occupied large plots, again sometimes in previously residential areas, reflecting the scale of their production.

Spatial "reach"

While the waterproof industry became locally embedded in the north of Manchester, it also had "reach" into larger-scale national and international economic flows. The waterproofing industry was for a long time dependent on natural rubber, which originally came from South America, then the Belgian Congo and later plantations in Southeast Asia. The story of how this rubber came to reach the city is as troubling as that of cotton, given the brutality with which workers were treated on colonial plantations [2]. On the other hand, the waterproofing industry was supported by a "reaching in" of skilled people to the city at the turn of the century, when Jewish immigrants from eastern Europe settled in the area of Cheetham Hill, just to the north of Victoria Station where they had first arrived into the city.

A supportive spatial culture

The Jewish community which emerged in Cheetham Hill hosted a strong "spatial culture" which is still discoverable in material culture and business routines today. Many members of the community started to specialise in a form of waterproofing called "schmearing". Because they did not have enough capital to set up large factories, schmearing often happened in homes, producing cramped, smelly and often dangerous working environments.[2] Garments were passed back and forth between machines, makers and shmearers in a multitude of local interactions, incorporating drying time between each process [6]. The life of the schmearers is vividly recounted in a series of novels by Maisie Mosco, who based her account on the experiences of a family who arrived in the city in 1904.[3] Mosco describes a local sharing of ideas and mutual support which was embedded in family ties but also enabled by local encounters at the market, in shops and at the synagogue. The shops and bakeries of Bury High Road were particularly important sites of knowledge sharing and labour market matching. This road (the left-hand fork of the city's foreground network which goes up past Strangeways) would also have effectively connected the area back into the rest of Greater Manchester.[4]

The rise of this more informal small-scale waterproofing activity created an industry that was overall marked by strong market segmentation, with some parts of the sector being dominated by fast fashion and others producing high-quality trade-marked garments in large-scale production processes in factories. Mandelberg & Co and Frankenstein & Sons, for example, registered a large number of patents for their designs [3]. These larger factories often employed a large number of women, and developed their own internal support communities, with Frankenstein & Sons, for example, regularly taking their staff on days out to Blackpool.

Interestingly, the Jewish waterproofing industry has left its mark on the Strangeways area that we have already explored as the centre of today's "fast fashion" cluster. Discussions on a Manchester-based internet forum reveal how during the 1950s and 1960s this area was a 'hive of activity' with 'its own magic'. A number of the buildings housing waterproofing that were in Strangeways in the 1950s still stand today, creating useful warehouse space for the fast-fashion clothing and accessories wholesalers. The Jewish spatial culture can also be found embedded in local routines, with many of these wholesalers retaining Saturday closures, associated with the Jewish Sabbath.

A tale of two cities: the outdoor leisure trade

As other parts of the textile industry declined, the waterproofing industry kept reinventing itself in the later 20th century, while exploiting the redundant machines and buildings left by the broader decline, including large roller and frame machinery left behind in empty mills – a 'spatially-fixed technology' which produced an unintended path dependency in the city [12].

A new line of production also emerged in the 1960s due to a "reaching out" across the Pennines to Sheffield [8]. The knowledge and capabilities embedded within both Manchester and Sheffield were brought together to produce innovative kit for outdoor leisure activities in the Pennines, where workers from both cities had long enjoyed walking and climbing. A community of expertise arose, focusing on the development of new, customised products which combined advanced fabrics and specialised metalwork – from lightweight rucksacks to detachable climbing nuts. From Greater Manchester's point of view, this meant producing fabrics that were waterproof, strong and light. From Sheffield's perspective, it required creating small and specialised forms of metalwork. Given that many of the participants in this community of expertise were themselves climbers, this represented a good example of "user-led innovation" based on an understanding of both 'what is needed' and 'what is possible' [12, p. 63]. While the community of practice started as a niche group, it generated a rich vein of innovation that has created companies that prosper to this day, in many cases becoming household names – such as Karrimor, Peter Storm, Mountain Equipment and Regatta.

In summary

This chapter used a case study from the waterproofing industry to illustrate processes of path interdependency and economic branching in Greater Manchester. The spatial context for this industry was identified to be both locally rooted and clustered, and internationally dependent in terms of both raw materials and

104   *Rebuilding Urban Complexity*

markets. As this sector branched out over time, it again and again produced new complementary economic footholds or "diversification surfaces", even in the context of general decline in the textiles and clothing industries. Overall it provides a rich example of what Jane Jacobs would identify as the development of "new jobs from old" [13], and the multiscale spatial processes that underly this development.

Notes

1 Just before the same process was discovered by Charles Goodyear in the United States.
2 People working in these conditions endured chemical poisoning and fires.
3 The Almonds and Raisins series was written between 1979 and 1991.
4 Laura Vaughan, a Bartlett-based professor of urban form who based her PhD thesis on the Jewish communities in Manchester and Leeds, points out that while this area of the city was sometimes characterised as a Jewish ghetto, it was locally spatially integrated and relatively well connected through the foreground network to the rest of the city.

References

1. Martin, R. and P. Sunley, Path dependence and regional economic evolution. *Journal of Economic Geography*, 2006. **6**(4): p. 395–437.
2. Jones, K. and P. Allen, Historical development of the world rubber industry. In *Developments in crop science, vol. 23. Natural rubber: biology, cultivation and technology*, M. Sethuraj and N. Matthew, Editors. 1992, Amsterdam: Elsevier. p. 1–25.
3. Parsons, M. and M.B. Rose, The neglected legacy of Lancashire cotton: industrial clusters an the U.K. outdoor trade 1960–1990. *Enterprise and Society*, 2005. **6**: p. 682–709.
4. Tadmor, Z. and C.G. Gogos, *Principles of polymer processing*. 2013, New York: John Wiley & Sons.
5. Schumpeter, J., *The theory of economic development*. 1934, Boston, MA: President and Friends of Harvard College.
6. Levitt, S., Manchester mackintoshes: a history of the rubberized garment trade in Manchester. *Textiles History*, 1986. **17**.
7. Harris, P., *Salford at work: people and industries through the years*. 2018, Stroud: Amberley Publishing.
8. Froy, F., *'A marvellous order': how spatial and economic configurations interact to produce agglomeration economies in Greater Manchester*. Bartlett School of Architecture. 2021, London: University of London.
9. Lancashire Life, *Private White VC – the designer label inspired by a Salford war hero*. 2014. https://www.greatbritishlife.co.uk/lifestyle/fashion/22620755.private-white-vc-designer-label-inspired-salford-war-hero/. Accessed [10/09/2024].
10. Clark, S., Chorlton Mills and their neighbours. *Industrial Archaeology Review*, 1978. **2**: p. 207–239.
11. Scott, P. and P. Walsh, New manufacturing plant formation, clustering and locational externalities in 1930s. *Britain Business History*, 2005. **47**: p. 190–218.
12. Rose, M.B., T. Love, and M. Parsons, Path-dependent foundation of global design-driven outdoor trade in the Northwest of England. *Information Journal of Design*, 2007. **1**: p. 57–68.
13. Jacobs, J., *The economy of cities*. 1970, New York: Vintage Books.

9 The destruction of urban complexity

Unfortunately, the *loss* of urban complexity is as important to the history of the post-industrial cities discussed in this book as is its emergence. We have already touched on the loss of economic complexity in Chapter 6, considering the entropic decline in manufacturing common to all post-industrial cities, which has nevertheless been somewhat compensated for by other forms of emerging economic complexity, in the services industry and the knowledge-based economy. This decline reflected, to some extent, the precarious nature of urban economies as "partially open" systems, where changes to global markets and competition from elsewhere undermined the viability of the complex manufacturing-based economies which had evolved in each place. However, also common to these post-industrial cities is a loss of *spatial* complexity, whose cause is very different – rooted as it is not in global market forces but rather in a very intentional set of policies and plans.

Victorian cities: sites of chaos and disorder?

The relatively coherent emergence of the street networks of Manchester, Sheffield, Newcastle and New Haven during their industrial heyday is a good example of how self-organisation can lead to a functional and resilient spatial organisation. Nevertheless, these Victorian cities were also characterised by squalor and poverty, as the exploitative practices of the larger factory owners spilled out into the city streets and as overcrowding led to the very "disurban" conditions associated with poor sanitation, pollution and lack of safety. Kim Dovey and colleagues document how such problems also develop in some (but by no means all) informal settlements in Global South cities today when either overcrowding or the speed of growth means that the positive characteristics of self-organised informal settlement are overridden by the development of 'slum conditions' [1]. The intermingling of industry and residential land uses, while positive in many ways, also created problems due to the pollution then associated with many industrial forms of production. This fuelled a set of commentaries which portrayed Victorian cities as sites of *disorder* as opposed to sites of emergent order [2]. These were then used to justify a set of planning policies that were orientated towards radical change.

A key writer who has influenced our impression of Victorian cities is Frederick Engels, whose family were mill owners in Manchester. In *The Condition of the*

DOI: 10.4324/9781003349990-12

106   *Rebuilding Urban Complexity*

Working Class in England, Engels decried the city as being an unplanned 'outgrowth of accident'. He portrayed Manchester and its surrounding towns as being 'labyrinthian', 'ruinous' and 'repellent' [3], with the area adjacent to the river Irk being a 'planless, knotted chaos of houses'. Engels in fact recognised the importance of the foreground network of busy commercial streets which existed in Manchester and Salford at this time, which, as he pointed out, hosted an 'almost unbroken series of shops'. However, he suggested that these commercial streets represented an intentional act of 'artifice' by the capitalist classes to disguise the poverty that existed behind them – describing how the 'aristocracy can take the shortest road through the middle of all the labouring districts to their places of business without ever seeing that they are in the midst of the grimy misery that lurks to the right and the left' [3, p. 58]. He thus ignored the more positive role that these streets would also have played in connecting the people living near them into the wider city. Similar descriptions exist of other industrial cities – indeed, the Crofts area of Sheffield was described in 1899 as creating 'feelings of repugnance and even horror' [4]. In 1910, in the United States, the New Haven Civic Improvement Committee identified poor conditions in dense working-class neighbourhoods, constituting 'back tenements, unsanitary shacks, crowding, secrecy and filth' [5, p. 136].

A new order?

Analysis of the historic street systems of these once industrial cities indicates that the underlying problems included poorly constructed buildings, a lack of sanitary provisions and a lack of pollution controls – not the structure of the underlying street system itself, which continued to successfully knit together these cities. These issues – which were associated with very low life expectancy – started to be addressed incrementally in Greater Manchester from the 1880s onwards [6]. Nevertheless, perceptions about the disorder and overcrowding of industrial cities still justified a series of top-down planning interventions in the mid-20th century. There was a new zeal for planning changes after the Second World War as part of the broader progressive social reform movement and the increasing use of the motorcar. This was an era in which architecture and town planning was in its element, with brave new plans being generated to impose new sorts of order on the city. Le Corbusier, for example, proposed that 'the city is crumbling, it cannot last much longer; its time is past. It is too old' [7, p. 4]. He developed radical new urban designs to accommodate the fact that the motorcar was placing unforeseen demands on city street systems. Indeed, Douglas Rae points out that while certain transport technologies, such as the tram, had been 'sustaining technologies' [5, p. 228] that enhanced street systems in cities like New Haven, conversely, cars were largely transformative, displacing people from the centre of cities to the suburbs and acting as 'decentering technologies'. English industrial cities also saw a flight to the suburbs in the early 20th century, which meant that planners had to take into consideration long-distance commutes. This was accompanied by a simultaneous urge to use urban spatial planning as a mechanism for tackling economic decline and the significant loss of local manufacturing jobs.

From streets to "walkways in the sky"

This broad post-war enthusiasm for sweeping urban change was to have a major disruptive effect on the spatial organisation of the centres of Manchester, Sheffield and Newcastle (not to mention other northern cities such as Coventry), helping to destroy part of the underlying complexity and integrity which had sustained these cities over the centuries.

In the Manchester city region, very few planning laws had been implemented before the 20th century, aside from some authorisations for street widening [8]. However, from 1945 onwards a comprehensive vision was set out to reconfigure the city. These plans were strongly orientated towards accommodating the car, reflecting concerns that central streets had become overloaded with traffic. However, a broader set of changes was also envisaged, following guidance from the Lord Mayor of Manchester, Leonard B. Cox, that the city replan its urban environment 'boldly and comprehensively . . . with no preconceived ideas or prejudices' [9].

The resulting 1945 plan for the city featured a set of ring roads (see Figure 9.1) which would be partly implemented in the 1960s, when they were incorporated in

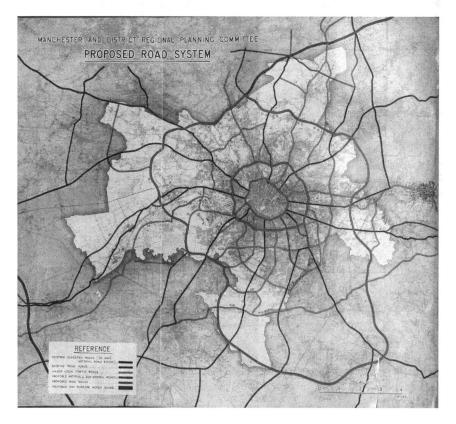

Figure 9.1 Map of proposed ring roads in 1945

Source: Nicholas, R.J., Manchester and District. Regional planning proposals. 1945, Jarrold: Norwich

108 Rebuilding Urban Complexity

a 1967 City Centre plan. The 1967 plan was based on a "distribution hierarchy" which also envisaged separating people from traffic through the creation of an 'elevated pedestrian landscape' [9, p. 18]. This thinking was strongly influenced by Colin Buchanan's 1963 *Traffic in Towns* report, which advised separating pedestrians and cars, and which informed, for example, the design of the long-imagined "Educational Precinct" on the Oxford Road. This precinct, originally envisaged in the 1940s, would have separated the city's universities from through movement (both north-south and east-west), while also elevating pedestrians to "walkways in the sky". The plans – which would have localised this knowledge hub from citywide knowledge exchange – were fortunately only partly realised, but they resulted in a first-floor shopping centre along the Oxford Road which closed in the 1980s, having been 'stranded by its first-floor location, literally up in the air' [10, p. 191].

More comprehensively realised, however, was the "slum clearance" which in the 1960s was directed towards what was identified as the 'planless, knotted chaos of dark, dismal and crumbling homes'[1] surrounding the city centre. This clearance included large areas of terraced housing in working-class areas such as Ancoats and Hulme [10], with "griddy" streets being largely replaced with more "cul-de-sac" style developments which channelled people into residential niches as opposed to providing them with broader choices of local movement. New social housing estates were also constructed, which were based on a very different concept to "street-based urbanism". They comprised high-rise buildings set amid large areas of open space, cut off from the surrounding urban fabric. Indeed, some of the social housing estates constructed at this time are now notorious, such as the Hulme redevelopment in Manchester and the Park Hill estate in Sheffield. Both fell well outside Hillier's set of spatial "constraints" underlying liveable forms of urbanism.

A set of images from the 1990s City Challenge regeneration in Hulme, for example, reveals how the old urban grain was destroyed in this area of Manchester to make way for new high-rise buildings arranged in "crescents" in open space (see Figure 9.2). The only historic buildings left standing were the corner pubs (see Figure 9.3). Thirty thousand homes were demolished across 230 acres of land [11].

Julienne Hanson has described the very particular morphology of the "estate-based" configurations which were built in their place. Housing estates tend to restrict the number of people in public spaces – not only because they isolate people from wider patterns of urban through movement, but also because they limit the number of entrances and active frontages onto public spaces. This absence of people in turn creates a negative feedback loop, as a lack of natural surveillance (Jane Jacob's "eyes from the street") creates a fear of going out amongst some groups, which is compounded when this same lack of natural surveillance encourages new uses of this space for antisocial and criminal behaviour [12]. Bill Hillier described the associated problem of "perpetual night syndrome" [13] – a situation where such housing estates end up having the same number of people in their public spaces during the day as would be expected in other, more coherent parts of the urban fabric by night.

The destruction of urban complexity 109

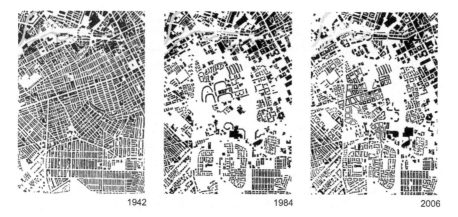

Figure 9.2 Changes to the Hulme urban fabric from 1942 to 2006

Source: These maps were hand-drawn as part of the Hulme City Challenge street-based redevelopment in the 1990s, provided to the author by David Rudlin

Figure 9.3 The Junction Pub in Hulme, 1984
Source: Richard Watt @McrHistory, www.mdmarchive.co.uk

Hulme's woes were increased by the closure to traffic of Stretford Road, which had been the area's principal shopping street, and severance from the city centre by the Mancunian Way, which '*ripped through*' Salford and Hulme [14], forming part of the city's "inner-relief route", whose construction started in the 1960s. (It was finally completed in 2004.) Such ring roads are very different from the longer

110 Rebuilding Urban Complexity

streets which connect cities in organically grown foreground networks (of which Stretford Road formed part). To start with, they are often elevated above ground and are mono-use, in that they carry only vehicular and not pedestrian traffic, and they often lack the intersections which knit roads and streets back into the local urban fabric. Their aim, to support rapid movement at a much larger – regional – scale, means that they lack the multiscalar connections which would help them to be properly *urban*. Indeed, it was fighting against just such inner-urban expressways in New York City that led Jane Jacobs to pit herself against the planner Robert Moses and mount a successful, and now famous, campaign of local community resistance.[2]

In 1973, community resistance in Manchester finally halted the long-held plans for the city's ring roads, following a public enquiry about the proposed demolition of a set of valued city-centre Victorian warehouses [9]. Nevertheless, the anticipation of the building of these roads had already influenced development and building design in the city centre for over thirty years, with one new building being designed, for example, to provide a sleek new façade which would 'reflect the gleaming metal of streamlined motor cars gliding past, at hitherto unknown speeds' (R Brook quoted in [9, p. 43]).

With regard to Hulme, the buildings which were left vacant following the failure of the new housing estates[3] provided, somewhat ironically, a form of urban "redundancy" that prompted a new flourishing of economic activity. The crescents became squats for a community of artists and musicians who used the area as a base for launching new streams of creativity that became pivotal to Manchester's reinvention as a centre for music and youth culture in the 1980s and 1990s, associated with Tony Wilson, Factory Records, the Hacienda and bands such as Happy Mondays and Simply Red [15] – a good example of the "self-organisation" which will emerge from inefficiency and redundancy in the city if allowed. When the area was finally regenerated in the 1990s – through the much-lauded City Challenge regeneration programme which reinvented a street-based urbanism[4] – this was felt as a loss by some [11]. The redevelopment followed a Hulme design guide which ensured that the area was developed through a system of connected streets instead of estates and cul-de-sacs [see 16].

Back in 1978, a large area of city centre complexity in Greater Manchester had also been destroyed to make way for the construction of the Arndale Shopping Centre, which was responsible for 'erasing the grain of the Victorian city in favour of a single monumental form' [17, p. 74]. The new centre directly destroyed a busy set of shopping streets around Market Street (as we have seen, a long-standing hub of the foreground network). At the same time, the indoor shopping space of nearly 93,000 square metres sucked away trade from neighbouring areas such as Oldham Street [10]. In 1996, an IRA bomb destroyed a substantial part of the centre, and this inadvertently provided an opportunity for a return to more street-based urbanism in this part of the city, leading to the development of the new interconnected Millennium Quarter, designed through international competition, which focused on 'reconnecting key elements to the core' [18, p. 170].

Equally dramatic planning interventions occurred in the complex, organically evolved, cities of Sheffield and Newcastle during the 20th century, which had both

Figure 9.4 Park Hill Flats in Sheffield
Source: Photo by Benjamin Elliott on Unsplash (https://unsplash.com/)

similarities and differences to Manchester's experience. Sheffield had been particularly hard hit by extensive air raids during the Second World War, giving a further impetus to top-down replanning efforts. In 1945 the document '*Sheffield Replanned*' set out a set of principles for a fifty-year schedule of works, which again strongly featured ring roads, and the Civic Circle Road was finally completed around the town centre in the 1960s. Again, pedestrians and cars were separated, with pedestrians being encouraged to use underground passageways ("subways"), especially at roundabouts. One such roundabout became known as the infamous "hole in the road", hosting a circle of below-ground shops accessible via escalators. As in Manchester, the aim was not to build on what had come before but rather to introduce 'radical changes in the patterns of urban living'.

The city also experienced a clearance of its terraced streets and the development of new estates comprising high-rise buildings set in open space – including Park Hill (see Figure 9.4) and others such as the Woodside/Burngreave Estate. At the same time, new developments on the periphery followed more of a rural rather than urban form, incorporating gates and lodges to create early forms of the "gated community". Meadowhall, a vast new shopping mall, was created, this time outside the city centre.

Initially, 'Sheffield was as proud of its new roads as of its housing, its clean air, and its flourishing arts. They were all symbols of rebirth after years of stagnation

112 *Rebuilding Urban Complexity*

among the ruins of the Industrial Revolution' [19]. However, the ultimate failure of these bold experiments promoted a long period of self-questioning, followed by more informed urban design based on historic patterns of "legibility". This process has been supported by an enduring Council-based urban design team that keeps things in-house as opposed to outsourcing expertise to consultants (as we will explore further in Chapter 11). In the contemporary city, pedestrians and traffic are no longer separated, and the pedestrian subway system of underground walkways has been closed and filled. Nevertheless, the city is still strangled by its inner ring road, with large roundabouts occupying areas which could have been used as civic spaces like the University Square. Sheffield is not alone in having such a legacy – ring roads continue to separate urban cores from their urban fabric in numerous other British towns and cities, including Reading, where I grew up. Concerns in Sheffield also continue as to the extent to which Meadowhall takes commerce away from the city centre; these concerns have led to a series of intensive city-centre retail schemes.

The city of Newcastle is also continuing to live through the impacts of urban planning changes which occurred in the 1960s and 1970s. We have seen that Newcastle experienced a relatively sympathetic reconfiguration of its city centre in the Georgian era, which created a sense of integrity at the heart of today's city, while connecting it to the original hub of lively commerce at the quayside of the River Tyne. The railways which came into Newcastle during the mid-19th century in turn also respected the existing urban fabric, entering the city on a viaduct which still works well with the surrounding urban fabric near the quay (see Figure 9.5).

However, the planning interventions of the 1960s and 1970s were much less sensitive, destroying swathes of urban complexity [20]. Again, this was largely intentional, representing an attempt to 'abolish the past and manufacture the future' (J.G. Davies [21], quoted in [20, p. 207]). The Leader of the City Council cited Brasilia, the newly planned capital of Brazil, as his reference point – an apt comparison, as Brasilia was also based on a top-down symbolic order which bore little resemblance to tried-and-tested forms of traditional urban morphology. Three sides of the Georgian Eldon Square (where my family's florist shop was located) were destroyed to make way for a large shopping centre similar to the Arndale Centre and Meadowhall. New roundabouts and roads cut off access to the quayside from the centre, with pedestrians being guided through (what remain) intimidating underpasses. The streets that the railway had not severed were at this point severed by the motorcar (see Figure 9.6).

These urban interventions were again accompanied by modern buildings which failed to have any relation with the street, a condition replicated to this day by glass-fronted examples of 'shining machined' buildings [9] which are currently being funded using UK Government Levelling Up funds, revealing we have learnt little from the past about the importance of engaged and porous frontages in producing lively streets. Indeed, as will be discussed further in Chapter 11, it seems difficult these days to escape the type of out-of-scale, sealed-off architecture which Douglas Rae described in New Haven as bearing 'every appearance of hostility to the smallness, the unevenness, the very humanity of a place' [5, p. 6].

The destruction of urban complexity 113

Figure 9.5 The railway viaduct coming through Newcastle quayside
Source: Photo by the author

In some respects, however, the city of New Haven itself had a lucky escape. Following the damning 1910 report by the Civic Improvement Committee which highlighted the disorder of the industrial city, a City Beautiful plan was drawn up, incorporating "Haussmann-style"[5] avenues. The plan was, however, rejected by Mayor Frank J. Rice (1869–1917), who insisted on more incremental changes such as improving the quality of sidewalks [5]. Douglas Rae celebrates this mayor, indeed, for '*not* transforming the city' (5, p. 105, my emphasis).

The impact of planning changes on industry

Economic activity in all three of these cities was also affected by the wider planning changes, which both coincided with and followed the broader period of manufacturing decline. One of the principal impacts was the forced separation of industry from other land uses in the city. In the United States, zoning had been introduced at the turn of the 20th century, with the first municipal zoning ordinance being introduced in Los Angeles in 1904. Douglas Rae describes how New Haven

Figure 9.6 Disurbanism in the heart of Newcastle
Source: Photo by the author

adopted zoning in 1926 – homogenising land uses and driving a wedge between commercial and residential areas [5]. By 1932, 766 U.S. municipalities had comprehensive zoning policies. These polices were granted legal backing following a Supreme Court case in Euclid, Cleveland in 1922. The outcome was to 'shift urban spaces away from heterogeneity of uses' [5, p. 262]. Indeed, the widespread impact of zoning led Christopher Alexander to pen a famous article in the 1960s, '*The city is not a tree*', where he argued against the idea that cities can be separated (or "branched") into different parts, each of which had different functions. He argued that cities were instead like lattices, constituting overlapping networks which cannot be easily separated into parts – again reiterating the importance of respecting the intricate part-whole structures of cities [22].

In the United Kingdom, zoning was not formally introduced in the same way as in the United States, but a complex set of planning regulations nevertheless enforced a separation of uses, and this had devastating effects on the intermingling of different economic activities which had made up the urban fabric of the country's previously industrial cities. While Greater Manchester, for example, is often imagined as having peaked in the industrial revolution, it was still very much an

The destruction of urban complexity   115

industrial city in 1950. As the city grew, the number of industrial plots grew at a faster rate than did the street segments.[6] In 1959, over half of the workforce was still employed in manufacturing [18]. However, many small firms were forced to either close or move out from more central locations as a result of the city's slum clearance and new planning processes (see the company history in Box 8). Rodgers [14] counts a thousand manufacturing firms lost in inner areas between 1966 and 1972, at least partly due to planning interventions. Indeed, a key aim of the 1945 plan drawn up by the City of Manchester was to 'unravel the conflicting mixture of land uses in the inner city' [23, p. 71], with planners asking themselves how could this 'intricate mix of uses – decayed residential, small-scale industrial and semi-derelict land abandoned by the transport industries – be teased out into a rational pattern?' [14, p. 29].

Box 8   Jay Trim – a firm that has adapted to urban spatial change

Haberdashery has long been a specialism in Manchester, with the city being known for the manufacture and sale of "smallware" (such as ribbons, threads, silk tapes and laces) in the 18th century [24]. Jay Trim is an example of the continuity of this capability. A family-run wholesale haberdashery based in Cheetham Hill that was founded in 1973, the company supplies a variety of customers across the UK, from manufacturers to retailers. The history of the firm tells a story of economic diversification and resilience in the face of urban upheaval and change. The current owner's father once worked in the electrical industry, but he had always wanted to run his own business. Inspired by the fact that his wife was a keen knitter, he started to sell wool (probably sourced from a spinners in Bradford) out of a suitcase at Salford market at the weekend. Eventually he ran two wool shops, one of which he was forced to close as a direct result of urban regeneration going on in Salford. While he kept his other shop, this failed to trade well as the area significantly lost population following the associated housing clearances. By the time new housing had been built, he had moved into trading in "seconds" and clearance lines associated with inefficiencies in the industry (where in the case of ribbon manufacture, for example, 100 metres of every 1,000 metres made would be substandard). He began by renting the top floor of a terraced house which had been adapted for commercial use a few hundred yards away from where the firm is based now. After a brief spell on the other side of Manchester, he returned to Cheetham Hill as this was 'the area where it was all happening'. The current owner first intended to train as an engineer but ended up in the family firm, alongside his mother and sister. He pointed out that while Strangeways was a textile-focused area it had always also supported economic diversity – being a base for wholesalers hardware,

116 *Rebuilding Urban Complexity*

> toys, fancy goods, mobile phone and other electronics. He feared, however, that competition from e-commerce could signal another shift in the area's primary economic activity.
>
> *Source:* The information in this Box is based on an interview carried out during my PhD research – see [29].

There was also an important impetus at this time to move firms into new segregated areas of the city – industrial estates and business parks – in part because of dangers to the health of people living close to the more polluting industries. Indeed, Greater Manchester was one of the very first cities to host an industrial estate. The Trafford Industrial Park was built on a former park to the west of the city in 1896 [10], and it influenced the construction of other such estates around the country via the 1909 UK Planning Act [25]. The industrial park, like its many copies across the world, shares many spatial characteristics with postwar social housing estates in that it is inward facing, with few connections into the surrounding urban fabric, and arranged in hierarchical or "tree-like" forms that create dead ends (see Figure 4.2 in Chapter 4). These effects are compounded by the lack of density and scale. As Howard Davis points out, industrial parks are often 'large in all dimensions, making their interiors distant from their edges' [26, p. 215]. This makes the park very intimidating to pedestrians. While many roads in the park are named streets and avenues, they are disurban in scale and lack active frontages. In fact, the industrial park is better connected regionally and nationally than it is locally, and hence has become home today to a large number of wholesale, distribution and logistics firms, taking advantage of the estate's strategic position in the UK as a whole. While this is economically beneficial to the city, the park represents a very large pocket of spatially segregated land very close to the city centre.

One reason that the park is now less attractive to manufacturing firms – particularly smaller ones – is that such areas cut off firms from the valued interdependences associated with urban agglomeration economies. To this day, many local policy makers continue to assume that manufacturing firms can easily be displaced to more peripheral locations in estates and parks – without a recognition that they too benefit from the broader interdependences which make up cities, particularly when firms are first starting out [27, 28]. Despite such practices, however, a surprising number of small manufacturing firms still base themselves in the central areas of cities such as Manchester and Sheffield [29, 30]. In London, work by Jess Ferm and colleagues has also revealed a considerable number of small manufacturing firms operating outside formal industrial zones in the city [31] – a sign again that not all forms of self-organisation can be quashed.

A loss of network integrity?

What has been the effect of these radical disruptions on the *configurational* spatial complexity of the affected post-industrial cities? Has there been an underlying *network effect* which might have implications for other dimensions of their social and economic complexity? The fact that the street systems of Manchester, Sheffield, Newcastle and indeed New Haven had largely retained their coherent structure and integrity until the 1960s is testament to the resilience of street systems. Julienne Hanson argues, for example, that 'the very grid itself may constitute an accumulation of strong morphological events which, taken together, produce a globally strong structure which is highly inertial and difficult to erase or destroy by local changes' [32, p. 186–187].

However, might these post-war planning changes have been of sufficient scale to reduce the 'organisational capacity' of the spatial networks underlying each city? [33, p. 148]. Closer analysis of the evolving street network of Greater Manchester suggests that this may indeed be the case. The structuring capacity of the city's foreground network seems to have weakened considerably since the 1950s, while the background network of the city also appears to have become more fragmented. These changes are made visible through the recent introduction of normalised variables in space syntax research, which support comparison between different cities (over space) and between the same city (at different points in time).

The 1850s urban area of Manchester and Salford had a very strong network of streets in comparison to other cities, which would have effectively pulled people into the city centre from surrounding towns. Indeed, the foreground network would have been much stronger in orientating such "through movement" than that of Sheffield at this time, which had a weaker foreground network but a more coherent background one (this finding for Sheffield is perhaps not surprising for a city which was less focused on trade and more centred around local artisanship, with artisans requiring connectivity to each other right across the urban fabric).

Until the 1950s, Greater Manchester retained a relatively strong foreground network, despite the city having expanded significantly by this time and at some speed. Although the foreground network of streets lost some of its power, the background network of the city became more cohesive by this time. In comparison, the contemporary map shows a much weaker structuring network, particularly for pedestrians, who are excluded from the mono-use parts of the network associated with inner-city expressways. A comparison of contemporary Greater Manchester with other world cities reinforces the idea that it has lost out to other cities in terms of the relative strength of its underlying spatial network – see [29]. In fact the centre of contemporary Greater Manchester (within the M60) has a lack of coherence in its background network relatively similar to that of Venice – a city which has the advantage of also being knitted together by its extensive canal system [34]. Greater Manchester is not alone in losing its coherence over time, with space syntax analysis revealing, for example, that Detroit, Michigan, had also lost an important amount of the underlying network structure which 'once served to build the interconnected city of industrial manufacturing' [35, p. 259]. Detroit's loss of coherence is much more evident and dramatic – with this post-industrial city being cited by local policy makers in Greater Manchester as having a fate which their own city has been lucky to escape[7]. Nevertheless, Sophia Psarra et al. point out that much of the unravelling of Detroit's spatial coherence has been small-scale and insidious. Such changes – which have also happened in Greater

118   *Rebuilding Urban Complexity*

Manchester – easily slip beneath the radar of policy makers, their social and economic impact going unrecognised.

Benefits from regional and national connectivity?

While Greater Manchester appears to have lost some of its internal coherence over time, it nevertheless remains relatively well-knitted into regional and national transport networks. Indeed, Greater Manchester outperforms other "core" UK cities[8] – Birmingham, Bristol, Leeds, Liverpool, Newcastle, Nottingham and Sheffield – in terms of its regional and national connectivity [29]. As we have already seen, this large-scale connectivity is clearly helping the city's warehousing, wholesaling and logistics companies. To some extent, the city is still very much a "warehouse town" [6]. Local policy makers in Manchester celebrate this regional and national connectivity, while prioritising new sites for economic investment (particularly in manufacturing) at nationally connected motorway junctions [27, p. 6]. However, while the city may be well connected at this larger scale, it may be less effective at bringing together economic capabilities at the local and city scale, a factor of importance if the city is to continue its transition towards being a hub of the knowledge-based economy.

Voids remain in left-over industrial land

While loss of structure can happen to any city, a particular problem affects post-industrial cities: the legacy of formerly industrial land. The loss of network coherence in Greater Manchester is made worse by the "voids" in the urban fabric left by once-industrial spaces (now designated as "brownfield land") creating an inner-city ring which separates the city centre from the rest of the conurbation. Sheffield similarly hosts areas of post-industrial void – for example in the Crofts area, which retained a mixed industrial and residential character until well into the 20th century, before much of the land was cleared to provide space for new industrial buildings which never materialised.[9]

Some of the land once occupied by factories and warehouses in Greater Manchester has been converted into large surface car parks, which not only reinforce car dependency but also create open spaces where once there was interconnected urban fabric. They in fact represented a key diversification strategy for former mill owners, who saw the car parks as a new business investment, with one interviewee in Bolton explaining to me that for every large car park and supermarket there was once a mill. Approximately 30,000 off-street car parking spaces exist in Manchester city centre, many of which are run by private companies such as APCOA, Citipark, Euro Car Parks, NCP Q-Park and SIP [36]. This represents nearly 5 per cent of the land within the Alan Turing Way ring road, compared with less than 3 per cent in Liverpool and Leeds [29].[10]

Manchester does, nevertheless, offer some very positive examples of how former industrial land can be renovated and revitalised, such as the canal-based Castlefield development to the southwest of the centre, and MediaCity, which has

The destruction of urban complexity 119

taken over the old Salford Quays, attracting many new broadcasting jobs to this part of the city, including from institutions like the BBC. Castlefield has become a tourist attraction and hub for nightlife, yet MediaCity remains relatively isolated from the city centre, meaning that it may be relatively inward-facing when it comes to knowledge exchange (in strong contrast to the Oxford Road Corridor, where the BBC used to be based). Sheffield has equally attempted to renovate a former industrial area since the 1980s through the development of a new cultural quarter, but has struggled harder to create any bottom-up self-organisation in an area which is again not well-connected into the rest of the city – a problem facing any city which is attempting to use culture as a tool for post-industrial regeneration, despite the positive examples associated with the Baltic and the Glasshouse International Centre for Music in Gateshead, across the Tyne from Newcastle.

Building "up" instead of "out"

The empty spaces which exist in Greater Manchester's inner-city ring could be prime spots for the introduction of new interconnected, street-based urbanism. However, what has actually taken place in Greater Manchester in recent years has been more ad hoc, fuelled by a flurry of international property investment projects, as investors seek to capitalise on uplift from the lower than average land values which used to exist in the city centre [37]. Many new property development projects are based on creating "vertical density" in the form of high-rise blocks set in unstructured forms of open space (a process which will be familiar from the 1960s), doing little to recreate the spatially configured networked street densities of the past. In his 2023 book, *Manchester Unspun*, Andy Spinoza reports that thirty new buildings of more than twenty stories have been completed since 2010, with a further twenty-three under construction [11]. Indeed, a group of academics at the University of Manchester [38] has gone as far as suggesting that policy makers are in the process of constructing a disconnected "new town" of high-rise buildings in the centre, with little respect to the existing city. This point is nicely captured by a mural on an old pumphouse in Broughton, Salford, which makes reference to that famous local painter of terraced streets, L.S. Lowry (see Figure 9.7).

Loss of commercial building space for manufacturing and artisanship in the city core

While formerly industrial land lies vacant in the inner-city ring, a contrasting problem is a loss of commercial buildings in the city centre, reducing the affordable incubation spaces available to smaller manufacturing and artisanal firms. Instead, the former textile mills – those 'abandoned landmarks of Manchester's former industrial glory' [39, p. 10] – have recently been the target of high-end residential property development, forcing out the small-scale entrepreneurs that had made use of these buildings, such as the Asian knitwear firms described in Chapter 7. A survey of textile sites in Greater Manchester by Historic England

Figure 9.7 A mural on a pumphouse in Broughton, Salford
Source: © Copyright Glyn Baker and licensed for reuse under a Creative Commons License

suggested that at least half have been lost since the 1980s, with the organisation regretting the loss of these buildings which could again become 'engines of prosperity' [40].

An example of this process in action is Crusader Mill – a grade II listed mill which was built around 1830 on Chapeltown Street, a formerly industrial area at the back of Piccadilly Station. This mill has recently been converted into loft apartments by a developer called Capital & Centric with support from the city government, and queues of people were eager to see the new flats when they came on the market. The tenants forced out by this development included not only the knitwear factories but also nearly 100 artists who had been taking advantage of this cheap, "messy" form of space close to the heart of the city. The gentrification cycle of artists squatting formerly industrial buildings in more derelict areas of the city, while unwittingly making these buildings more attractive for residential development, is well known already from places such as Hackney Wick in London. In the case of Crusader Mill, the artists were actively supported to find new accommodation by Arts Council England, but the same help was not extended to the numerous knitwear manufacturers who had inhabited the building for over fifty years [see 41], preserving and maintaining the building during this time. One of the former resident artists, Sam Meech, has actively campaigned on behalf of these knitwear companies,[11] highlighting the ways that private developers selectively use industrial history in their marketing strategies [41]. As he says, while the property developers celebrated the old textile mill in social media, there were no references to the

The destruction of urban complexity 121

recent knitwear firms – 'no tweets about the giant Shima Seiki knitting machines that whirred on every floor; not even a single joke about Roy Cropper cardigans or bad Christmas jumpers'. He describes this as a wider example of how 'developers rapidly knit their own narrative, unravelling existing patterns and encoding their own motifs, row by row, tweet by tweet' [41, p. 21].

Opportunities have also been missed to set aside appropriate commercial spaces in the new property developments happening close to the city centre. The recent redevelopment of Ancoats, the area so key to the industrial revolution, provides a case in point.[12] Here old heritage mill buildings have been restored at the heart of a high-density residential project comprised of multistorey residential blocks which make little provision for artisanal or industrial commercial spaces, despite the likely "incubation" benefits associated with having these spaces close to the city centre. A very commercial decision was made to not integrate workshop spaces into the development, based on the difficulties of "pricing in" such spaces while making the rest of the development financially viable. While top-down planning may have been the key reason for a loss of urban complexity in the 1960s and 1970s, today it is more likely to be commercially led, profit-maximising property development which is problematic – as will be discussed further in Chapter 11.

In summary

This chapter has described how the underlying urban morphologies which had evolved and persisted in formerly industrial cities such as Manchester, Sheffield and Newcastle over centuries were disrupted by top-down planning changes in the mid-20th century. It also explored the specific impacts that these planning changes had on industry, destroying the intermingling of economic uses that had characterised these cities up until this time. Space syntax analysis of the changing spatial configuration of Manchester revealed how the city had lost *configurational structure* following these planning interventions – a fact that is not strongly appreciated by local policy makers. A disconnect between the city centre and its surrounding urban fabric is compounded by industrial voids of land which separate valuable new developments such as MediaCity from the city centre.

In the very last chapter of this book, we will reflect on how local policy makers can respond to these challenges, rebuilding urban complexity both spatially and economically. We will also explore the barriers faced by policy makers in seeking to rebuild complexity at a time when commercial property development has largely replaced bold public-sector master plans in shaping urban development. Before this, however, we will take a necessary detour, exploring a set of systems associated with the natural environment which have long been ignored in our economic and architectural narratives. While Victorian cities are often decried for their negative impacts on the environment, today we can learn from aspects of their economic and spatial complexity, as we attempt to rebuild urban complexity in a way which also respects the *natural* diversity and complexity on which it depends.

122 *Rebuilding Urban Complexity*

Notes

1 See the statement by Alfred Morris in Hansard: House of Commons Debates Volume 721, Monday 22 November 1965.
2 The fight between Jacobs and Moses is very effectively chronicled in the 2016 film *Citizen Jane: Battle for the City*.
3 A tenant survey in 1975 indicated that 96 per cent wanted to leave – cited in Andy Spinoza (2023), Manchester Unspun.
4 As can be seen from the image to the right in Figure 9.2.
5 Georges-Eugène (or Baron) Hassumann was responsible for a major urban renewal programme in Paris in the mid-19th century, demolishing large numbers of buildings and constructing new avenues, boulevards, and parks.
6 An analysis of historical maps reveals that while the size of the city increased 20-fold between 1850 and 1950, the number of street segments increased 4-fold, and the number of industrial plots marked on the historical maps increased 6-fold.
7 This view was expressed, for example, in discussions around the city's Independent Prosperity Review.
8 This term is used to describe a set of 11 important UK cities, London excluded. See www.corecities.com/cities. Accessed [07/06/2024].
9 See http://sytimescapes.org.uk/zones/sheffield/S10 which is part of a series of very useful maps and historic descriptions prepared by the South Yorkshire Historic Environment Characterisation project. Accessed [29/05/2024].
10 Based on the Parkulator shapefile provided by Dr Tom Forth – see https://github.com/thomasforth/parkulator. Accessed [07/06/2024].
11 Sam Meech created various artworks based on the Crusader Mill experience, including *Fait Accompli* (featuring knitted tiles that reproduced the peeling, hand-created signage used by the textiles companies), a series of "knitted portraits" of the people who used to run the factories who had been rendered invisible by the development process, and a film called *Social Fabric* (2019), which documents the displacement of one of the factories.
12 These observations resulted from participation in an Academy of Urbanism tour. www.academyofurbanism.org.uk/ancoats/. The views expressed here are the author's own and do not represent those of the Academy of Urbanism.

References

1. Dovey, K., et al., *Atlas of informal settlement: understanding self-organized urban design*. 2023, London: Bloomsbury Publishing.
2. Griffiths, S. and K. Navickas, The micro-geography of political meeting places in Manchester and Sheffield c. 1780–1850. In *Micro-geographies of the Western City, c. 1750–1900*. 2020, London: Routledge. p. 181–202.
3. Engels, F., *The condition of the working class in England*. 1993, Oxford: Oxford University Press.
4. Belford, P., Work, space, and power in an English industrial slum: "The Crofts," Sheffield, 1750–1850. *The Archaeology of Urban Landscapes: Explorations in Slumland*, 2001: p. 106–117.
5. Rae, D.W., *City: urbanism and its end*. 2005, New Haven: Yale University Press.
6. Groom, B., *Made in Manchester: a people's history of the city that shaped the modern world*. 2024, Dublin: Harper Collins.
7. Corbusier, L., *The city of tomorrow and its planning*. 1971, London: The Architectural Press.
8. Farrer, W. and J. Brownbill, The city and parish of Manchester: introduction. *The Victoria History of the County of Lancaster: Lancashire*, 2003. **4**: p. 2003–2006.
9. Brook, R. and M. Jarvis, *Trying to close the loop: post-war ring roads in Manchester*. Centre for Environment and Society Research Working Paper Series no. 24. 2013, Birmingham: Birmingham City University.

10. Wyke, T., B. Robson, and M. Dodge, *Manchester: mapping the city*. 2018, Edinburgh: Birlinn Ltd.
11. Spinoza, A., *Manchester Unspun: pop, property and power in the original modern city*. 2023, Manchester: Manchester University Press.
12. Hanson, J., Urban transformations: a history of design ideas. *Urban Design International*, 2000. **5**: p. 97–122.
13. Hillier, B., *Space is the machine: a configurational theory of architecture*. Vol. 2007 e-print. 1999, Cambridge: Cambridge University Press.
14. Rodgers, H.B., Manchester revisited: a profile of urban change. In *The continuing conurbation: change and development in Greater Manchester*, H.P. White, Editor. 1980, Gower: Farnborough. p. 26–36.
15. Wray, D.D., 'It was like blade runner meets Berlin rave': the Manchester sink estate with the UK's wildest nightclub. *The Guardian*. 2023, London.
16. Rudlin, D., The Hulme and Manchester design guides. *Built Environment (1978-)*, 1999. **25**: p. 317–324.
17. Canniffe, E., The morphology of the post-industrial city: the Manchester mill as 'symbolic form'. *Journal of Architecture and Urbanism*, 2015. **39**(1).
18. Peck, J. and K. Ward, *City of revolution: restructuring Manchester*. 2002, Manchester: Manchester University Press.
19. Waterhouse, R., Will Sheffield snarl up? *The Guardian*. 1972, Manchester.
20. Barke, M., B. Robson, and A. Champion, *Newcastle upon Tyne: mapping the city*. 2021, Edinburgh: Birlinn.
21. Davies, J.G., *The evangelistic bureaucratic: a study of a planning exercise in Newcastle upon Tyne*. 1972, London: Tavistock Publications.
22. Alexander, C., A city is not a tree. *Design Magazine*, 1966. **206**: p. 46–55.
23. Lloyd, P.E. and C.M. Mason, Manufacturing industry in the inner city: a case study of Greater Manchester. *The Royal Geographical Society*, 1978. **3**(1): p. 66–90.
24. Maw, P., Provincial merchants in eighteenth-century England: The 'great oaks' of Manchester. *The English Historical Review*, 2021. **136**(580): p. 568–618.
25. Historic England, *Industrial buildings: listing selection guide*. 2017, London: Historic England.
26. Davis, H., *Working cities: architecture, place and production*. 2020, Abingdon, Oxon and New York: Routledge.
27. New Economy, *Deep dive 0.2 manufacturing*. 2016, Manchester: Manchester University Press.
28. GMCA, *Greater Manchester's plan for homes, jobs and the environment: Greater Manchester spatial framework revised draft*. 2019, Manchester: GMCA.
29. Froy, F., *'A marvellous order': how spatial and economic configurations interact to produce agglomeration economies in Greater Manchester*. Bartlett School of Architecture. 2021, London: University of London.
30. Jones, M.J.J., *Sheffield at work: people and industries through the years*. 2018, Stroud: Amberley Publishing.
31. Ferm, J., D. Panayotopoulos-Tsiros, and S. Griffiths, Planning urban manufacturing, industrial building typologies, and built environments: lessons from inner London. *Urban Planning*, 2021. **6**(3S2): p. 350–368.
32. Hanson, J., *Order and structure in urban space: a morphological history of the City of London*. Bartlett School for Architecture and Planning. 1989, London: University of London.
33. Griffiths, S., Manufacturing innovation as spatial culture: Sheffield's cutlery circa 1750–1900. In *Cities and creativity from the renaissance to the present*, I.V. Damme, B. De Munck, and A. Miles, Editors. 2018, New York: Routledge Advances in Urban History. p. 127–153.
34. Hillier, B., T. Yang, and A. Turner, Normalising least angle choice in Depthmap and how it opens up new perspectives on the global and local analysis of space. *Journal of Space Syntax*, 2012. **3**(2): p. 155–193.

124   *Rebuilding Urban Complexity*

35. Psarra, S., C. Kickert, and A. Pluviano, Paradigm lost: industrial and post-industrial Detroit – an analysis of the street network and its social and economic dimensions from 1796 to the present. *Urban Design International*, 2013. **18**(4).
36. Robson, S., The end of driving into town? Almost half of Manchester's city centre car parks are set to disappear. *Manchester Evening News*. 26 September 2020, Manchester.
37. Lewis, C. and J. Symons, *Realising the city: urban ethnography in Manchester*. 2017. Manchester: Manchester University Press.
38. Folkman, P., et al., *Manchester transformed: why we need a reset of city region policy*. CRESC Public Interest Report. 2016, Manchester: CRESC Centre for Research on Socio-Cultural Change.
39. Werbner, P., Renewing an industrial past: British Pakistani entrepreneurship in Manchester. In *Migration: the Asian experience*, J.M. Brown and R. Foot, Editors. 1994, Basingstoke: Palgrave Macmillan.
40. Cushman & Wakefield, *Engines of prosperity: new uses for old mills. North West*. 2017, London: Historic England.
41. Meech, S., Fabrications: using knitted artworks to challenge developers' narratives of regeneration and recognise Manchester's South Asian working class textiles businesses. *Textile*, 2022: p. 1–22.

Part Three

10 Cities as systems of systems

Where nature fits in

So far in this book we have considered two particular examples of urban systems: economic and spatial. However, cities host many other interdependent systems, often being described as "systems of systems". We have already touched several times in this book on *social* systems – the set of overlapping social relationships in which urban economies are embedded, from friendships to family relationships to ethnic and religious ties. Indeed, these relationships are often crucial to the trust and goodwill which underlies economic collaboration and problem-solving [1]. This relationship also applies in reverse, with economic ties forming the basis for social relationships. Douglas Rae described how in New Haven, for example, multiple relationships of goodwill became established through the shops that proliferated in the city in its heyday: between shops and suppliers but also shops and their customers – suggesting that all local residents were ultimately 'wrapped up in this web of relationships, being a "somebody" as one produced, sold, bought, and consumed the stuff of everyday life' [2, p. 74]. Evolutionary economic geographers such as Ron Boschma have analysed the social factors which increase "proximity" between economic actors and hence increase the likelihood of collaboration – revealing that organisational, social and institutional forms of proximity are as important as geographic and cognitive forms in this respect [3, 4].

It is equally clear that not all networks are of the egalitarian, distributed kind which were so important to the establishment of the initial stages of the industrial revolution in Manchester or the metalworking cluster in Sheffield. Manuel DeLanda has documented the oscillation which has occurred in European history between more distributed and generative economic networks (which he calls "meshes") and more rigid hierarchical relationships (which work to reduce dynamism) [5]. This is clearly visible in the way that Manchester became at one point dominated by large cotton companies that restricted other forms of diversification in much the same way as Detroit became dominated by a few automobile firms. Hierarchical relationships very often come at a cost to the workers at the bottom of the pyramid – with Manchester's prosperity becoming increasingly polarising and unequal as the factory system developed. The exploitation of people which was central to the extraction of resources and raw materials at a global scale during the colonial era was also accompanied by a callous exploitation of workers at a much more local scale.

DOI: 10.4324/9781003349990-14

128   *Rebuilding Urban Complexity*

Likewise, while it is beyond the scope of this book to explore the *social movements* which have emerged in cities like Manchester – antiprotectionist, feminist, antislavery, trade-unionist, and others – these are as much a product of the agglomeration of people in a place as the economic interdependencies that we have focused on here – and equally influenced by local spatial configurations [see 6, 7].

What we will be focusing on mostly in this chapter, however, is another system which has a profound impact on cities but which has been long neglected: the natural ecosystem. Given increasing concerns about anthropogenic impacts on climate change and widespread biodiversity loss, it is indeed difficult not to think these days of *linked social-ecological systems* when imagining cities [8, 9]. Philip Sykas, for example, asked in 2022 'whether an innovation can be considered an advancement if it does not further environmental sustainability or social equity' [10, p. xxx].

This is all the more important for Manchester, as it is often held up as an example of a place where our relationships with nature started to go badly wrong – as was mentioned in Chapter 6, the "fossil capital" which kickstarted our economic dependence on fossil fuels and the era of mass-production stemming from the industrial revolution [11]. Seen through this perspective, the path interdependences which have so far been celebrated in this book start to take on a more negative light – particularly given that there are enduring psychological and societal tendences and biases (such as a disregard for nonimmediate consequences) which may make it challenging for us to break out of such path dependence and fully "problem-solve" our way towards a less damaging relationship with nature in the future.[1]

A lattice economy? The creative use of by-products in the Victorian city

How can a complex systems approach help us understand likely paths towards a greener future? To start with, we might benefit by looking backwards towards the complex webs of production which helped to *reuse* waste in the past, as opposed to simply dispose of it. While Victorian industrial cities did indeed begin a long-standing global dependence on fossil fuels, there were nevertheless good practices in these cities from which we could learn, not least in the treatment of industrial waste. Such practices made urban economies of this time more circular (compared with today's linear "take-make-waste" economies). They were based, however, on "branching" processes and cross-sector interdependences which might make the term "lattice economy" seem more appropriate as a description.

The economic geographer Pierre Desrochers has argued that Victorian cities were particularly efficient at reusing their waste and "loop-closing": exploiting, as far as possible, industrial by-products for new lines of innovation [12, 13]. This very often created cross-fertilisations between industries – in 1763 one commentator described the process 'of enriching one art with the practices, materials, and sometimes even the refuse-matters of another' (quoted in [10, p. xxiv]). Desrochers identified such practices as one of the important benefits of early forms of agglomeration. As time progressed, the reuse of waste began to be seen as a symbol

Cities as systems of systems: where nature fits in   129

of modernity – a 'measure of the degree of industrial development and capability' [10, p. xxx].

The textiles trade, so important to Manchester, has a particularly rich history of exploiting its waste products, where 'few opportunities for re-use were missed' [10, p. 2]. A Cotton Waste dealers directory in 1882, for example, included over 700 entries and twenty waste trade specialities. As the waste textiles industry itself diversified and branched, it often did so on the basis of different kinds of process and different types of machine. Some of these processes also required skilled work by hand – such as the sorting of rags into a 'myriad of fibre, colour and quality categories' [10, p. xxiv] to reduce the need for recolouring and dyeing. Cross-sector collaborations involving waste were also common between the textiles and paper industries. Rags were a key raw material for the early paper mills until the 1870s, with door-to-door collections of old clothes by "rag and bone men" to supply the mills. While clothing provided the necessary fibres, bones were collected to provide gelatine to seal the surface of the paper.[2]

In Victorian Manchester, frequent notices appeared calling for waste products and advertising that waste products were available for reuse. For example, in 1890, Edwin Butterworth and Co submitted the following notice: 'WANTED, New Vulcanised and Non-Vulcanised India-rubber Cuttings, Old Buffers, Rings, Piping, Goloshes, Tubes, Valves, &c. – Edwin Butterworth, Henry Street, Ancoats.[3] The company also dealt in cotton waste, engine-cleaning waste, hides and woollen rags, and wood pulp and paper-making materials. There was a *spatiality* to the recycling and waste trades in the city. For example, in the 1880s, cotton waste traders were concentrated in Manchester and Oldham, and at a more local scale in Bolton, Bury, Heywood, Rochdale and Stockport [10]. Later, analysis of the 1950s map of the city reveals a cluster of recycling firms to the north of the city which included scrap metal operators, repair shops (e.g., for shoes and cash machines), rag warehouses, cotton and textiles waste dealers, wastepaper works, general waste cleaning works and a furniture depository. Often residual spaces were used to store and process waste – including, for example, a disused church on the corner of King Street and Duke Street, which had become a metal scrapyard. The GOAD maps reveal other spaces available for storing and passing on materials; the map of the area around Dale Street (featured in Figures 7.3 and 7.4) shows various sheds and yards for storing rags and waste materials – a "spatial architecture" to support circularity and the reuse of waste products which is mostly missing from our city centres today.

The reuse of waste does not just reduce the impact of urban economies on their environment, of course – it also provides new "lines of flight" and innovation in the economy, harnessing 'human creativity, know-how and entrepreneurship to create wealth out of residuals' [14, p. 38]. This was again something anticipated by Jane Jacobs, who foresaw that waste recycling could play a key role in the future of urban economies, with cities likely to become 'mines of materials'. She pointed out that cities that take the lead in this process will be able to export not just new products but also the innovative *processes, technologies and equipment* used for making these products. Cities like Amsterdam are indeed today moving ahead with better monitoring of waste production and local material flows, and taking a

130   *Rebuilding Urban Complexity*

particularly proactive approach to circularity [15]. However, there are many challenges ahead.

One of these challenges is contamination. When it comes to waste recycling, it becomes obvious that while some systemic relationships between materials are "extrinsic" (i.e., the integrity of the parts are preserved as they come into relationship with other elements, so that they can easily be disassembled), others are "intrinsic" (i.e., the different elements merge in ways that are difficult to reverse) [16, 17]. Intrinsic relationships can create problems when it comes to recycling materials for reuse, as in the case of the blended textiles which have come to dominate the textiles and clothing industries. Some textiles experts in Greater Manchester speak nostalgically about the simpler, single-fibre products which were easier to recycle in the past – and which in fact generated greater complexity when it came to supply chains. Fine worsted suiting, for example, was often later recycled into chair upholstery and carpets. The contamination created by intrinsic relationships is also very familiar to post-industrial cities due to the difficulty of recycling and reusing *buildings and land* after they have been vacated by polluting industries (see Box 9).

Box 9   Contamination – when complexity becomes problematic

"Intrinsic" relationships between materials are more difficult to manage than extrinsic ones, and they remind us that relational complexity is associated not only with positive forms of interconnection but also with more disturbing relationships of contamination, pollution or indeed contagion – as we became only too aware during the Covid-19 pandemic. The contamination which causes problems when it comes to recycling and reuse is pointed out very effectively in Michael Braungart and William McDonogh's book, *Cradle to Cradle* [17].

One of the particular problems facing many post-industrial cities is land contamination. The land and buildings left over from the industrial era are expensive and difficult to reutilise due to the chemicals and industry pollutants which now infuse them. Tim Edensor of the University of Manchester refers to these remnants as "ruinous matter" [18] – describing the particular set of affordances characteristic of "ruined things" which lose their discreteness and challenge conventional forms of classification. Post-industrial cities can, however, exploit such challenges as sources of new problem-solving and innovation – the Materials Processing Institute in Middlesbrough, in the northeast of the UK, for example, has been developing new methods of decontaminating brownfield land, and is using these processes as a source of "urban mining". The "ruinous" sites in Manchester described by Edensor are also often sites of important biodiversity despite their pollution – hosting natural forms of self-organisation, and complex new ecological niches, as we will come back to later in this chapter.

Cities as systems of systems: where nature fits in   131

Another challenge is the availability of space in city centres for recycling, repair and reuse – given that cities like Manchester are already struggling to demarcate and preserve commercial spaces for manufacture and artisanship. While some recycling processes are best kept peripheral due to their polluting emissions, others (such as repairing, sharing and repurposing) rely on the same agglomeration economies that benefit manufacturing firms. One opportunity for such "reproductive" processes may be the empty spaces which have emerged on many high streets and retail centres as conventional retail stores become outcompeted by e-commerce. Nevertheless, the affordability of such spaces is important – recycled products produced in expensive city locations characterised by high rents are unlikely to be competitive or within the everyday financial reach of city dwellers. While local agglomeration economies will be important to the emergence of such "lattice economies" in the future, Pierre Desrochers also warns against too much "localism" when it comes to recycling and reuse, suggesting that in many circumstances it might be best to source a reused material from another, more distant agglomeration which has developed particular specialist capabilities and the economies of scale which accompany these. Again, it is important to remember that cities act as both partially open and also *complementary* economic systems.

Broader green industrial transitions

It is not just waste management which will need to evolve if we are to green our urban economies. The nature of production itself will also need to change. As identified earlier, path dependency is often seen in a negative light when it comes to the environmental impact of urban economies. Nevertheless, the economic branching processes which have been so important to the evolution of cities like Manchester and Sheffield may equally be important in the transition towards greener futures. There has been a commitment by many city governments to transition to "net zero" – that is, a point at which no additional carbon emissions are produced through the way people live, work and consume (by 2038 for Greater Manchester, and 2040 for Sheffield, for example). While policy makers in Greater Manchester are largely focusing on reducing the carbon emissions of the city's building stock, in Sheffield, there is more of a focus on greener industrial transitions – building on and adapting the capabilities that are already present in the city. One policy document argues that

> the same capabilities that put the City Region at the heart of the world's first industrial revolution can put us at the centre of the fourth – producing new materials, new processes, and new answers to the environmental, social and wellbeing challenges facing the UK and the world.
>
> [19, p. 7]

In order to develop this greener future, Sheffield policy makers are focusing in part on "clean steel", whilst also exploring new sectors such as hydrogen.

132   *Rebuilding Urban Complexity*

A literature review carried out with colleagues on green industrial transitions in 2023 found that a degree of path independency will indeed be important to this process [20]. Local economies that host embedded *capabilities* that are related to "greener industries" appear to make the transition toward greener production more easily – even if the previous industries that they have hosted are "brown" (i.e., polluting) rather than green [21]. Indeed, the economic geographer Phil Cooke has long pointed to the role played by underlying industrial capabilities in the development of green technologies. He showed, for example, how the world-leading wind turbine technologies which now exist in the region of Jutland in Denmark developed from fan-based technologies that had been used in milk cooling when the region was predominantly agricultural. These same rotary blade technologies later became used for marine engine-cooling, before becoming a key technology in windmill production [22]. In the UK, renewable wind energy development, carbon capture and materials recycling is happening particularly rapidly in Teesside and Middlesbrough, in the northeast of England, with new jobs in these sectors offering promise to a region that had long fought to make up for the loss of jobs in the coal, steel, and chemicals industries in the late 20th century. Some researchers are helping to predict the "green adjacent possible" for local economies, comparing regions in Mexico, for example, to identify what green products they might diversify into, given their existing capabilities [23]. More random economic diversity also remains important to radical green forms of innovation, however, and there is also interesting research into how local economies in the Global South could surpass the slower transitions towards net zero which are taking places in more complex economies, through, for example, the earlier use of renewables.

All this work reveals that mapping the interconnected economic "landscape" of a city (as was done in Chapter 3) can help us to better understand how a city's current constellation of sectors might either constrain or enable future greener development paths. However, cities commiting to "net zero" will also need to be mindful of products that may come into their cities from distant, less regulated economies.

An urban fabric built for walking and cycling

The focus in this chapter has so far been on urban economies. However, street systems are also important to the environmental impact of cities in myriad different ways. One strong focus in contemporary urban policy is the *walkability* of urban environments – not only for environmental but also for health reasons. Active travel (including walking and cycling) has been shown to have an array of positive health outcomes [see, for example, 24]. However, a recent study of 794 cities across the world found that more than half of respondents still commuted to work by car every day, and the authors cited Manchester as an example of a city where more than two out of three journeys are by car [25]. It is now clear that the planning changes which at first sought to *accommodate* the demands of the motor car in cities have since made these same places car-dependent. The current mayor of Greater Manchester, Andy Burnham, stated in a seminar in 2020 that the city had been built for the car 'for a long, long time',[4] with this leading to a wide set of

Cities as systems of systems: where nature fits in 133

social costs – not least air pollution. It is estimated that exposure to the current level of small particulates in the air in Greater Manchester contributes to around 1,200 early deaths per annum [26].

While "walkability" is now a widespread goal, there is less awareness, however, of how the spatial configuration of street systems will make this an easier goal to achieve in some cities rather than others. A study by the company Space Syntax found that there is an important correlation between the spatial configuration of street systems and the degree to which people actually walk in towns and cities across the UK. Physical infrastructural change may be needed to make "walkability" a reality in some places. The study found that higher levels of walking were more likely in bigger cities. However, there was a city size effect, with other active modes such as cycling and public transport replacing walking as a means of getting to work for many once a city reached a certain physical size [27]. London, for example, has a very high use of public transport and active travel commuting as compared with other UK cities. Bill Hillier was keen to point out that such organically grown cities will always offer a more ecologically sensitive and less polluting way of living than the "eco-villages" which continue to be proposed on the periphery of cities and towns – where people often still rely on their cars to get to jobs and services.

An idea closely related to the "walkable city" is that of the "15-minute city", which became popular post-pandemic based on work by Carlos Moreno at the Sorbonne [28] and an initiative put forward by the mayor of Paris, Ann Hidalgo.[5] This concept – that people should be able to meet most of their everyday needs within a short period of time – has become somewhat controversial in the UK [29]. Nevertheless, policy makers in Melbourne, Australia, have estimated that ensuring people can access everyday amenities within a similar 20-minute range would reduce travel by nine million passenger kilometres, cutting Melbourne's daily greenhouse gas emissions by more than 370,000 tonnes.[6] Again, a *configurational* understanding of street networks is necessary if such an idea is to work. Parts of cities that already provide access to multiple amenities at a 15, 20 or even 5-minute radius (such as much of central Paris) are generally an *emergent* product of the multiscale properties of the street system – which also provides people with access to more distant opportunities across the city. The shutters are likely to go up quickly on commercial amenities (shops, bars, cafés) that are "placed" in neighbourhoods where this multiscale layering of accessibility is not working. The concept has more relevance, however, for public services (public transport, health centres, child care, local hospitals, libraries) where the public sector *can* plan where things should go and also factor in the maintenance of such services for the longer term.

Returning biodiversity to cities

The complex characteristics of street systems also need to be taken into consideration as policy makers consider how to best bring biodiversity back into cities. While urban cityscapes have long been seen as the opposite of rural biodiverse

134 *Rebuilding Urban Complexity*

landscapes, in fact cities have recently been identified as a "safe haven" for diversity, given the monoculture/single species agriculture which exists across much farmed countryside. The Greater Manchester Combined Authority is taking this seriously, for example, with a five-year Environment Plan, in addition to specific schemes such as its Tree and Woodland Strategy. Manchester, Newcastle and Sheffield have transformed their relationships with their rivers and canals in recent years, bringing nature back into the city via these once polluted routes so important to industry. Indeed, the Oxford-based economist Dieter Helm[7] identifies that we will need to start *governing* at the scale of river basins in the future to better manage water-based ecosystems, pollution and flooding – yet another example of the *level* at which you intervene in a system being important to how you achieve meaningful and holistic outcomes [30].

While the benefits of bringing greater biodiversity into cities are clear, bringing new forms of complexity into already complex urban systems can be challenging.[8] Trees, for example, introduce shade but also potentially reduce the "natural surveillance" which is so important to feelings of urban safety. While concern about trees and street safety can be exaggerated, care evidently needs to be taken not to disrupt the close relationship between 'eyes on the street' and 'eyes from the street' which Jane Jacobs identified as being so important to keeping city streets safe. In Manchester, the "City of Trees" initiative (a project which dates back to Michael Heseltine's time as environment secretary) focuses on planting ornamental varieties which have canopies that are sufficiently high so as not to disrupt lines of vision. The City of Trees initiative has also helped to highlight the relationship between trees and other urban systems through a comprehensive mapping exercise – identifying every single site in the city where a tree could be planted (i.e., areas greater than 7 square metres which are not designated or protected sites) and then scoring each tree according to the advantages that it might bring in the context of other systems, such as surface water management, air quality, proximity to areas of social deprivation and linkages into biodiverse corridors – a good example of an approach which recognises the true complexity of cities as "systems of systems".[9]

As referred to in Box 9, interestingly, some argue that many vacant brownfield sites should *not* be developed because of the complex biodiversity that they have come to host since their abandonment by industry. While one argument emerging from this book is that such sites should be redeveloped as complex areas of urban grain to knit the city back together, academics at the Manchester Metropolitan University point to the fact that over half of this land in Greater Manchester hosts vegetation, offering hidden pockets of green space. They have developed a typology of twenty-six different types of brownfield land in the city to assist policy makers in deciding which sites to develop and which sites to leave alone or convert to nature-based community use. A good example of the latter approach is Nutsford Vale in Gorton, which was first a quarry, then a landfill site, and is now a biodiverse habitat developed by the local community into an area accommodating an orchard, wildflower meadow and plenty of pathways for walkers.

Cities as systems of systems: where nature fits in   135

Nature-based infrastructures

Policy makers in post-industrial cities such as Manchester are also starting to look at the potentials associated with nature-based infrastructures – again, looking beyond the simple and clear value of having nature back in the city for its own sake to understand the potential for interaction with other urban systems. This reflects a broader re-conception of nature as incorporating valuable "green infrastructures" which may help tackle the climate-related challenges which cities are now facing. Authors such as James J. Kay extol the benefits of harnessing the *self-organising* characteristics of nature-based systems for human purposes. This means that bringing nature back into cities will again require an understanding of the *systemic* properties of both natural and human-made systems to better predict how they might interact. Manchester City Council (the municipality this time) now has a green and blue infrastructure plan, which includes "Sponge City" thinking, an idea originally from China which seeks to use natural systems of permeability to support flood control. The West Gorton Sponge Park was set up in 2020, for example, to mop up water from extreme weather events.

In harnessing green and blue infrastructure, post-industrial cities can also learn from elsewhere in the Global South. In cities such as Kolkata, in West Bengal, natural systems are already playing an important role as local infrastructures. There is currently strong research interest in a human-made wetland system to the east of this city, which has over time become a multipurpose ecological, economic and social infrastructure while also treating over half of the city's sewage [31]. The Melbourne-based design thinker Dan Hill favourably compares such green infrastructure to more costly purpose-built high-tech infrastructures such as the Gates Foundation's "omniprocessor" in Dakar, Senegal.[10] The wetland system of Kolkata was developed in the 1940s when the city's sewage was routed along waterways into the Bay of Bengal. Local farmers exploited these waterways and ponds for fish production, and the fish, algae and water plants became a natural self-organising system to purify the sewage, helped by sunlight and wind. The wetlands now host considerable biodiversity, while supporting diverse economic livelihoods – eliminating the need for fertilisers for agriculture and vegetable production. The wetlands also act as a toxic sink for methane, carbon, nitrogen and sulphur while providing flood protection for the city. They have been recognised as a Ramsar site – a wetland of international importance.

Unfortunately, this example of a natural green infrastructure also offers us a reminder that the self-organising properties of systems need to be valued in order to be maintained. The wetlands area has shrunk significantly in recent years, being encroached on by informal and illegal property development [32]. Dhrabajyoti Ghosh, an environmentalist and engineer who dedicated his life to the preservation of the wetlands, noted that because no political leader in Kolkata can take credit for the wetlands system (which developed "bottom up"), there has been nobody to defend this infrastructure [see 31]. As Dan Hill points out, a key problem with infrastructure in general is that it is "backstage" and taken for granted, even if the opportunity costs of not having the infrastructure will be financially serious. This

136 *Rebuilding Urban Complexity*

example also reminds us that in order to properly consider post-industrial cities as future *socio-ecological systems*, we need to think at a city-region scale, considering how cities work with their immediate natural hinterlands.

In summary

This chapter has taken a broader perspective on urban complexity in post-industrial cities, going beyond the systems embedded in urban economies and the built environment to also look at the ecological systems with which they interact. It has argued that while cities such as Greater Manchester have been presented in a negative light due to their role as "fossil capitals", Victorian industrial cities also incorporated a strong ethos of "reuse", leading to a lattice economy which we could learn from today. While path dependence is often seen as a negative aspect of our relationship to nature, in fact the economic branching so important to industrial cities in the past may also be important to how they become greener in the future. The chapter also explored the importance of the spatial configuration of street systems to walkability and health, and considered how nature and biodiversity can be brought back into the city while respecting the existing complex systems that lead to lively urbanity.

In the next and final chapter of the book, we refer back to these socio-ecological considerations as we consider more broadly how local policy makers in post-industrial cities can *rebuild urban complexity*, fighting back against damaging planning processes – both historic and contemporary.

Notes

1 See, for, example Patrick H. Byrne's rich 2003 article 'Ecology, economy and redemption as dynamic', which discusses the role of problem-solving and emergence in the work of Jane Jacobs and Bernard Lonergan, while also identifying the psychological biases which can limit processes of problem-solving and intelligent "self-correction", particularly in relation to natural-human environments.
2 See www.manchesterhistory.org/reprints/PaperMillsHerald.pdf and http://ivybridge-heritage.org/archive/the-use-of-rags-in-paper-making/. Accessed [07/06/2024].
3 Grace's Guide to British Industrial History. www.gracesguide.co.uk/. Accessed [07/06/2024], with advert originally appearing in the *Derbyshire Advertiser and Journal*, 7 September 1890.
4 This view was expressed at an event where policy makers and economists revisited the city's Independent Prosperity Review in the light of the Covid-19 pandemic. See www.resolutionfoundation.org/events/city-limits-covid-19/. Accessed [07/06/2024].
5 www.paris.fr/dossiers/paris-ville-du-quart-d-heure-ou-le-pari-de-la-proximite-37. Accessed [07/06/2024].
6 Such an idea obviously has social benefits as well as environmental ones. This type of thinking gained particular traction after the Covid-19 pandemic, when many people relied on local amenities as a way of avoiding public transport and keeping out of busy city centres. This concept has also been welcomed by authors who argue that female perspectives need to be brought more into planning and design, such as May East in her 2024 book *What If Women Designed the City* – particularly as women are often "time poor" due to caring responsibilities.

Cities as systems of systems: where nature fits in   137

7 Dieter Helm influenced national environmental planning via his chairing of the National Capital Committee from 2012 to 2020.
8 With thanks to Nicolas Palominos for our many interesting conversations on this topic, and also to Peter Stringer for filling me in on all the activities which have been going on to increase tree coverage in Greater Manchester.
9 https://mappinggm.org.uk/. Accessed [07/06/2024].
10 https://medium.com/slowdown-papers/11-post-traumatic-urbanism-and-radical-indigenism-c2a21dc7ba69. Accessed [07/06/2024].

References

1. Ikeda, S., The meaning of 'social capital' as it relates to the market process. *The Review of Austrian Economics*, 2008. **21**: p. 167–182.
2. Rae, D.W., *City: urbanism and its end*. 2005, New Haven: Yale University Press.
3. Boschma, R., Proximity and innovation: a critical assessment. *Regional Studies*, 2005. **39**(1): p. 61–74.
4. Torre, A. and D. Gallaud, *Handbook of proximity relations*. 2022, Cheltenham: Edward Elgar Publishing.
5. DeLanda, M., *A thousand years of nonlinear history*. 1992, New York: Zone Books.
6. Navickas, K., *Protest and the politics of space 1789–1848*. 2016, Manchester: Manchester University Press.
7. Pinarbasi, S., Manchester antislavery, 1792–1807. *Slavery & Abolition*, 2020. **41**(2): p. 349–376.
8. Kay, J.J., An introduction to systems thinking. In *The ecosystem approach: complexity, uncertainty, and managing for sustainability*, D. Waltner-Toews, J.J. Kay and N.E. Lister, Editors. Chichester: Columbia University Press, p. 3–13.
9. Dixon, T.J. and M. Tewdwr-Jones, Urban futures: planning for city foresight and city visions. In *Urban futures*. 2021, Bristol: Policy Press. p. 1–16.
10. Sykas, P., *Pathways in the nineteenth-century British textile industry [Kindle edition]*. 2022, Abingdon, Oxon: Routledge.
11. Malm, A., *Fossil capital: the rise of steam power and the roots of global warming*. 2016, London: Verso Books.
12. Desrochers, P., Does the invisible hand have a green thumb? Incentives, linkages, and the creation of wealth out of industrial waste in Victorian England. *Geographical Journal*, 2009. **175**(1): p. 3–16.
13. Desrochers, P., Bringing interregional linkages back in: industrial symbiosis, international trade and the emergence of the synthetic dyes industry in the late 19th century. *Progress in Industrial Ecology, an International Journal*, 2008. **5**(5–6): p. 465–481.
14. Desrochers, P., Market processes and the closing of 'industrial loops': a historical reappraisal. *Journal of Industrial Ecology*, 2000. **4**(1): p. 29–43.
15. Williams, J., *Circular cities: a revolution in urban sustainability*. 2021, London: Routledge.
16. DeLanda, M., *Intensive science and virtual philosophy*. 2002, London and New York: Continuum.
17. Braungart, M. and W. McDonough, *Cradle to cradle*. 2009, New York: Random House.
18. Edensor, T., Waste matter – the debris of industrial ruins and the disordering of the material world. *Journal of Material Culture*, 2005. **10**(3): p. 311–332.
19. Sheffield City Region, *The Sheffield city region renewal action plan*. 2020, Sheffield: Sheffield City Region.
20. Froy, F., et al., What drives the creation of green jobs, products and technologies in cities and regions? Insights from recent research on green industrial transitions. *Local Economy*, 2022. **37**(7): p. 584–601.

138 *Rebuilding Urban Complexity*

21. Mazzei, J., T. Rughi, and M.E. Virgillito, Knowing brown and inventing green? Incremental and radical innovative activities in the automotive sector. *Industry and Innovation*, 2023: p. 1–40.
22. Cooke, P., Transversality and transition: green innovation and new regional path creation. In *Path dependence and new path creation in renewable energy technologies*. 2016, London: Routledge. p. 89–106.
23. Pérez-Hernández, C.C., et al., Mapping the green product-space in Mexico: from capabilities to green opportunities. *Sustainability*, 2021. **13**(2): p. 945.
24. Kelly, P., et al., Systematic review and meta-analysis of reduction in all-cause mortality from walking and cycling and shape of dose response relationship. *International Journal of Behavioral Nutrition and Physical Activity*, 2014. **11**: p. 1–15.
25. Prieto-Curiel, R. and J.P. Ospina, The ABC of mobility. *Environment International*, 2024. **185**: p. 108541.
26. Transport for Greater Manchester, *Greater Manchester's clean air plan to tackle nitrogen dioxide exceedances at roadside*. 2022, Manchester: Transport for Greater Manchester.
27. Parham, E., E. Jones, and E. McCoshan, *Understanding how urban form enables walking*. 58th ISOCARP World Planning Congress. 2022, Brussels.
28. Moreno, C., *The 15-minute city: a solution to saving our time and our planet*. 2024, New York: John Wiley & Sons.
29. Horton, H., Why has the '15 minute city' taken off in Paris but become a controversial idea in the UK? *The Guardian*. 2024, London.
30. Helm, D., *Green and prosperous land: a blueprint for rescuing the British countryside*. 2019, London: HarperCollins.
31. Watson, J. and W. Davis, *Lo-TEK: design by radical indigenism*. 2019, Cologne and Los Angeles: Taschen.
32. Banerjee, S. and D. Dey, Eco-system complementarities and urban encroachment: a SWOT analysis of the East Kolkata Wetlands, India. *Cities and the Environment (CATE)*, 2017. **10**(1): p. 2.

11 Rebuilding urban complexity

How can policy makers intervene?

> "And if we weren't nearly blind to the property of self-organization, we would do better at encouraging, rather than destroying, the self-organizing capacities of the systems of which we are a part".
>
> – Donella Meadows [1, p. 79–80]

This book has explored how cities act as catalysts – intertwining economic capabilities to produce innovation and diversification. We have reflected on the importance of the *networked* spatial characteristics of street systems in allowing cities to play this catalytic role. Indeed, they have been shown to be a neglected factor in how cities persist and reproduce themselves as economic entities. We have learnt how spatial and economic proximities in cities combine to create an underlying *field of potential* which has the power to support creativity and resilience as cities evolve and adapt to new economic realities – and when this is destroyed, this undermines how cities function as "complex adaptive systems". This final chapter will summarise the key messages of this book before circling back to policy and considering the implications for local policy makers – the people who I was initially hoping to reach when embarking on this academic journey. In the process we will explore the systemic characteristics of policy making itself.

What have we learnt?

We have delved into urban complexity through a three-stage process – looking first at the part-whole relationships characteristic of systems, then at system evolution and finally at cities as "systems of systems". We explored the fact that spatial and economic networks create "part-whole" structures in cities – with streets hosting people and movement according to their position within the city as a whole and with individual economic sectors benefiting from multiple interdependences. By focusing on how economic communities are "expressed" in one particular city – Greater Manchester – it is possible to see how the particular industries that make up a city might be harnessing shared capabilities in the form of labour pools, embedded knowledge and shared production networks. The "foreground network" of streets was seen to help cities function as one "giant workshop", with industries profiting

DOI: 10.4324/9781003349990-15

140   *Rebuilding Urban Complexity*

from mutual accessibility, whatever their exact position in the urban fabric. Rather than cities being "objects" which are tightly bound in space, they were shown to be partially open systems, connecting into broader economic networks – with different parts of the city offering specific operational access points or *vantage points* into these networks.

In Part Two, we considered how the particular sets of industries that can be found in cities today are the result of historic *path interdependency* – a historical branching and diversification of industrial sectors which leaves its trace in contemporary economic structures in a form of "related variety". Qualitative "ancestry analysis" in Greater Manchester revealed how the textiles, clothing, chemicals and engineering industries have co-evolved and cross-fertilised since before the industrial revolution, with ongoing problem-solving processes leading to the emergence of "new work" from "old" [2]. When economic communities reach a "critical mass" of locally concentrated capabilities, this constitutes a local resource which keeps producing new lines of economic activity – such as the waterproofing industry which emerged from the 'textiles and clothing' community but which also brought in capabilities from the chemicals and industrial gas sectors, and later from metalworking in Sheffield. We saw that it is not just economic networks that evolve over time but also spatial networks. In northern cities such as Manchester, Newcastle and Sheffield, the incremental evolution of city streets created powerful urban structures which brought people into the heart of these cities and supported their social and economic interaction. The "deformed wheel" provided a supportive "backbone" through different periods of economic history. However, these configurational structures were disrupted through planning interventions from the 1960s onwards, in ways which may be having long-term repercussions for how these cities work as "agglomeration economies".

Finally, in Part Three we focused on how cities operate as "systems of systems", exploring the socio-ecological relationships which are increasingly fundamental to sustaining urban life today. We reflected on what we might learn from the "lattice" economies common to Victorian cities based on the reuse of industry by-products, while also considering how self-organising *natural* systems (such as the East Kolkata wetlands) could be better harnessed by cities in the future. In order for cities to become more sustainable, a new systemic approach is becoming necessary which knits economic, social, spatial and ecological systems together in more symbiotic ways.

Circling back to policy

What are the implications for local policy? There has been an increasing recognition of the need for policy making to evolve in the light of complexity, with ideas from evolutionary theory, systems and complexity theory influencing thinking on how policy is made and how it could be organised differently [see, for example, 3]. There is a growing recognition that policy makers can work most efficiently by *building on the latent potentials* of self-organising systems as opposed to designing things from scratch. UCL academic Mike Batty argues, for example, '[A]s we

Rebuilding urban complexity: how can policy makers intervene? 141

learn more about the functioning of such complex systems we will interfere less but in more appropriate ways' [4, p. 771], with David Chandler discussing the idea that, '[G]overnance needs to be reframed in order to recognise the creative and self-ordering power of life itself' [5, p. 62]. As illustrated by the quote from Donella Meadows which introduces this chapter, it is now recognised that policy makers could benefit from *harnessing* processes of self-organisation to achieve their goals, rather than either denying or impeding them.

However, in order to act in a context of complexity, policy makers first need to recognise that they themselves are embedded in their own complex systems. Back in the 1950s, Charles E. Lindblom published a famous paper titled 'The science of "muddling through"' [6], where he described how, in the face of complexity, most policy makers follow their own forms of limited path dependency. They generally only ever seek partial solutions to issues on the basis of marginal choices between policies that have already been tried (although not necessarily evaluated) in a process of successive limited comparisons. As mentioned in the Introduction, this description of policy making resonated with me after my own experiences of working in this sector. Interestingly, Lindblom identifies that the policy making process follows a branching-type process which will not be unfamiliar after reading this book. However, this branching process is relatively disconnected from the complex phenomena which policy makers are seeking to address.

To counteract this tendency, Horst Rittel and Melvin Weber argued in 1973 for a new approach in urban planning to tackling the "wicked" – that is, complex and systemic – problems which occur in cities. They advocated consulting widely with different interest groups in order to locate problems before intervening – 'finding where in the complex causal networks the trouble really lies' [7, p. 159]. They also referred to the challenge of deciding on the boundary of the system to be focused on – something which, as we have seen, is particularly difficult to determine in open, interacting urban systems. These authors did not necessarily argue for an incremental approach, pointing out that incremental changes can sometimes further embed path dependency, with the overall path not necessarily being the desirable solution to a given situation. Incremental changes can provide obstacles (lock-in) which prevent larger-scale valuable systems change.

From incremental intervention to the encroachment of "market-based rationalities"

Despite such recommendations, there has been a notable turning away from such radical interventions in *urban form* by urban planners today, with a renewed interest in the evolutionary approaches put forward by people like Patrick Geddes and Jane Jacobs, who both advocated incremental interventions rather than bold master plans [8–10]. Geddes called for 'conservative surgery' to rectify urban problems while also advocating 'survey before plan'[1] – suggesting that planners spend time observing and documenting how places work before imposing new designs. Jane Jacobs called equally for "unslumming" as a way of renovating and retrofitting neighbourhoods that had fallen into disrepair – empowering people to slowly

142 *Rebuilding Urban Complexity*

improve conditions in their neighbourhoods without destroying the urban grain that underlay them.[2] Jacobs railed against the planners who saw only disorder and chaos and sought to impose their own ideas of order, arguing that '[T]he trouble with paternalists is that they want to make impossibly profound changes, and they choose impossibly superficial means for doing so' [11, p. 354].

These ideas have been particularly influential in changing the response to informal settlements and situations of rapid urbanisation in Global South cities, where the approach has for a long time now been to avoid demolition, and rather set about a process of in situ "upgrading" (see, e.g., [12, 13]). Matias Echanove and Rahul Srivastava reflect on the continued relevance of Patrick Geddes's ideas, for example, for development processes in Mumbai, in India. They argue for 'an urban gaze and practice that takes Mumbai's existing forms and dynamics into account and builds on them rather than against them' and interventions that are crafted 'into the logic of a living settlement or neighbourhood' [14, p. 280, 287].

To some extent planning *theory* has also turned away from concerns of urban form, towards a preoccupation with the *procedural* aspects of "making plans" and achieving adequate community participation [15–17]. Where architects and urban planners have taken a step back from large-scale urban interventions, "market-based rationalities" [5] have stepped in to fill the gap, with many aspects of city centre redevelopment in both the United States and the United Kingdom now being managed through an uneasy alliance between city governments and property investors. In British cities a long period of laissez-faire neoliberal government has also led to an increasing reliance on market-based rationalities to "get urban development done", particularly in the context of wider budget austerity since the 2008 global economic recession. In the absence of adequate public funding, compromises are often made, based on a complex set of concerns about affordability, likely profitability and the possibilities for job safe-guarding or job creation. The concerns of "getting a particular site to work financially" often override any thought about how each new development will knit back into the wider urban fabric or contribute to the wider urban integrity of the city.

Tilting the playing field

One thinker who imagines a stronger role for public policy in the context of complexity is Mariana Mazzucato. Mazzucato has become a "celebrity economist" for her work with governments around the world, at both national and city level.[3] She recognises the power of self-organisation by both people and markets, but at the same time argues that there is a need to 'tilt the underlying playing field' so that there is an incentive for private actors to solve problems which are in the public interest [18]. She argues for a "missions-led" approach, which overcomes "bounded rationality" (which we will explore further later) to bring together people from different policy areas around a common objective. Transposed to the fields of urban design, architecture and planning, such thinking propels us to develop a vision for how we would like our cities to work and function, and helps to steer self-organisation in this direction. Recent calls for planners to return to "urban

Rebuilding urban complexity: how can policy makers intervene? 143

visions" despite the challenges of complexity are also relevant here (see, for example, the work of Timothy J. Dixon and Mark Tewdwr-Jones [19]).

Overcoming bounded rationality

A final systemic issue which is increasingly recognised in policy circles is that of "bounded rationality". This term was originally developed by Herbert Simon in 1957 to refer to limitations in economic decision making, but since then it has been used as a broader term to describe how people often "bracket off" wider concerns to focus on their own specialist area, with this limiting the realisation of "part-whole" relationships. Donella Meadows describes, for example, how '[T]the bounded rationality of each actor in a system – determined by the information, incentives, disincentives, goals, stresses, and constraints impinging on that actor – may or may not lead to decisions that further the welfare of the system as a whole' [1, p. 110]. This is particularly important given that developing a coherent approach to cities as "systems of systems" requires an important overlap and synergy between professional mindsets (see e.g. [20, 21]). For example, staff responsible for developing Greater Manchester's industrial strategy recognised that while there were spatial factors that might be influencing agglomeration economies in the city, their potential to influence change in this area was limited because the city's spatial framework was also being developed simultaneously in a different department. Sometimes, there is notable path interdependency between some sectors and a lack of interconnection between others. For example, architects are more likely to work with engineers than anthropologists – despite the clear importance of buildings in providing a context for social interaction [see also 22].

New policy directions in the context of complexity and emergence

Bearing in mind these complexities and the systemic characteristics of policy, this final section of the book sets out new policy directions which may help to rebuild urban complexity. While these new directions are aimed at policy makers in post-industrial cities, they will be of wider relevance to all those working to create more sustainable, creative and spatially integrated cities. In the process, examples will be brought in from towns and cities across the UK – not just Sheffield and Newcastle but also London, Birmingham and the seaside town of Folkestone.

Fourteen new policy directions for post-industrial cities

Tackle "relic" infrastructure: While a Geddes-style approach to planning would suggest preserving and enhancing what is already there through "conservative surgery", more radical strategies will be needed to finally remove the "relic" post-war infrastructure – the mono-use ring roads, expressways and large roundabouts – which so disrupt the urban fabric in the heart of many post-industrial cities. Such an approach has worked relatively successfully in places such as Seoul (with the removal of an elevated highway making space for the Cheonggyecheon stream

144   *Rebuilding Urban Complexity*

restoration) and Boston (with the Big Dig project that rerouted another elevated interstate highway into the O'Neill Tunnel).

Reconnect urban cores to peripheries: New areas of street-based urbanism will also be needed – particularly where there are "voids" in the urban fabric left by the decline of industry. Rather than building *up* (a dominant development mode in many cities today), cities with old industrial inner-city rings need new urban development to connect *outwards* from their cores if they are to reconnect these cores to the wider city and become less car-dependent.

Learn "systems lessons" from history: In developing new areas of urban fabric, planners and architects need to learn from the "organically grown" urban form of historic cities. Such learning does not require a nostalgic return to traditional aesthetic designs, or even the ad hoc adoption of "what works" principles from historic villages (as has occurred in places such as Poundbury, outside Dorchester, influenced by the architectural theorist Leon Krier). Instead, it requires an interrogation of the *systemic* characteristics of what has worked in the past, taking into account the multiple scales of connectivity which organically grown street systems support.

Jane Jacobs criticised, for example, the New Urbanist school in the United States, for trying to re-create mixed-use neighbourhoods without recognising the underlying urban anatomy and the crossings and convergences of street networks which are required for local economic activities to survive. As she argued in an interview in 2001,

> Big cities have a lot of main squares where the action is, and which will be the most valuable for stores and that kind of thing. . . . But they're always where there's a crossing or a convergence [. . .] You can't stop a hub from developing in such a place. You can't make it develop if you don't have such a place. And I don't think the New Urbanists understand this kind of thing. They think you just put it where you want.
>
> [23]

This idea that you can just "place economic activities" in cities is also an important undercurrent in the discourse around the 15-minute city, as discussed in Chapter 10.

Do not overcomplicate the solutions: Rebuilding spatial complexity does not itself need to be complex. While theorists such as Luis Bettencourt suggest the need for a technical "science of cities" approach to "designing for complexity" [9], we have seen in this book that the key ingredients which nurture complexity in organically grown cities are actually quite simple. Bill Hillier and his colleagues set out relatively few basic morphological principles (or "constraints") which allow cities to work as lively urban spaces, such as the "deformed wheel" or foreground network of streets which are so important to knitting together cities into wholes. When intervening and developing new urban fabric at the local scale, planners and architects need to be mindful of this foreground network, linking new parts of the city back into it to ensure multiscale connectivity.

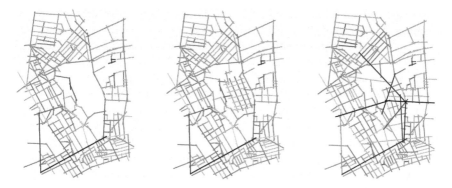

Figure 11.1 Three different ways of knitting the urban fabric back together
Source: Space Syntax (https://spacesyntax.com/)

Create grids that are connected back into the foreground network: Analysis of the historic environments of Manchester, Sheffield and Newcastle also revealed the importance of *grids* – confirming the statement by architect Leslie Martin that grids "afford complexity". Grids distribute movement, allowing people choice as they move through the street network, and promote encounter at their many intersections [24]. Nevertheless, rebuilding grids needs to be done with care. While sometimes it will be sufficient to build a relatively uniform grid across a void of space in the urban fabric, in other cases paying attention to the surrounding foreground network will be key. An example from a bid to redesign the area behind Kings Cross site by Norman Foster and the UCL spin-off Space Syntax serves as an illustration. The first diagram in Figure 11.1 shows the void north of the station which needed to be filled. The second diagram shows how this void could be filled just with a simple grid pattern. The third diagram, however, reveals via space syntax analysis that properly connecting the area back into the broader city would require a set of oblique *longer streets* set at angles to connect into the surrounding foreground network.[4]

Create streets, not roads: Another key learning from organically grown cities is the importance of "streets" as opposed to "roads" in creating urban vitality and safety. Streets have a specifically urban quality which is associated with buildings fronting more directly onto them, with multiple windows and doors offering an "active frontage". Indeed, Bill Hillier argued that for streets to be safe, it was important that there were lines of dwellings on both sides. The "Create Street" movement in the UK has recently had some success in highlighting the essential contribution that streets bring to urban environments. Entrances and windows offer the "eyes on the street" and the natural surveillance that Jacobs felt was so important, with this being further supported when street-facing businesses such as cafés and shops are accommodated. Because streets are linear in form, they provide connection into the wider city through long lines of sight, with this more global connectivity bringing the complementary "eyes *from* the street" which are the final

146   *Rebuilding Urban Complexity*

key component of network-based urban vitality. Such street-based urbanism can be combined with relatively high densities – as evidenced by Manchester's remaining city core with its griddy network of five- or six-storey warehouses and mills that have active frontages. Indeed, this core (which is depicted on this book's cover) is not that dissimilar to the type of multistorey street-facing residential and commercial buildings that dominate successful, walkable European cities such as Paris and Hamburg. Greater Manchester's policy makers could do well to extend this "integration core" of the city where they can, extending its continuous street-facing density into the formerly mixed industrial ring which surrounds it. This would mean taking over poorly utilised land, such as surface-area car parks, while also using brownfield land in ways which are sensitive to the complex biodiversity that these "left-over" areas of the city have come to support, and taking advantage of their potential to harbour biodiversity and offer green space in the heart of the city.

Make buildings and their facades permeable: This book has focused mainly on the "syntax" or grammar of the built environment – the way in which buildings are arranged in space and what this means in terms of the networked flow of public space in cities. It has generally not considered the semantics of buildings – their aesthetic "look" or scale. Nevertheless, building frontages matter when it comes to their interplay with streets and hence their ability to support human-scale, liveable environments. Too often new property development in UK cities still results in the "blank facade" – already ubiquitous on the outside of shopping malls or cinema complexes in urban centres.[5] Just as problematic are the glass facades now used so often for office construction. These facades may appear transparent, but they often have very few entrances, creating a visual idea of openness (of sorts) but no real interaction between the street and the people inside the building. As Richard Sennett identifies, such buildings create *boundaries* rather than borders – constituting to some extent 'sealed glass boxes'. Sennett asserts that "making buildings more porous will be one of the great challenges of 21st Century architecture" [25, p. 47].

Introduce design codes, not master plans, and learn from "pop-up" urbanism: The redevelopment of new areas of street-based urban fabric in cities should be carried out through hands-off design codes or guidance frameworks, as opposed to top down master-plans based on "one single intention" [26]. In developing such design frameworks, contemporary policy makers and planners can learn from the incremental approach which was in fact taken in Sheffield in the 1990s, inspired by Richard Roger's *Urban Task Force* report and the consequent Urban Design Compendium published by English Partnerships and the Housing Corporation. A design team embedded in Sheffield Council developed a local version of this compendium which set out "degrees of intervention" in different parts of their city, identifying areas which could benefit from incremental "evolutionary" change and then areas which would require bigger changes, while still following character-based "quarter guidance". The compendium took a broad view of design, with an important focus on the legibility of the urban environment, and the possibilities for maintaining traditional street configurations in new developments, with this appreciation for spatial configuration and "legibility" continuing to some extent in the city today.[6]

Rebuilding urban complexity: how can policy makers intervene?   147

The compendium was a forerunner to the current national interest in "design codes" which set out principles and priorities for local urban design to help guide multiple bottom-up interventions. While design codes are currently being piloted in only a few urban areas, having such a code in place might have helped to avoid a recent public outcry in the seaside town of Folkestone, on the South Coast, following proposals for a new high-density, harbourside development which many felt was out of keeping with the character of the rest of the town [27] [see Box 10].

Box 10   Lessons not learnt from "pop-up" developments in Folkestone

Recent proposals for a new high-density, harbourside development in Folkestone, on the South Coast came as a particular shock to local residents as the harbour area had previously seen more sensitive regeneration. This has included a flourishing of low-rise "pop-up" style shops and cafés along the harbour arm, associated with the renovation of the historic railway station (from which people used to embark onto ferries), and a transformation of the old railway track into a flower-lined walkway back to the rest of the harbour [28]. This latter regeneration scheme has provided a site of bottom-up innovation which builds on local heritage, while also sympathetically connecting back to the town through both physical pathways and views across to the more traditional parts of the harbour and its fishing boats. Such small-scale interventions and associated temporary "niches of complexity" provide indications of what works which are then ignored as profit-maximising logics take over in larger, more permanent developments.

Bring economic diversity back to city centres: The blank facades previously discussed represent a double assault on complex urbanism – not only do they remove "eyes on the street", they also fail to offer niches for commercial activities and small-scale entrepreneurs. Another key lesson that we can learn from organically grown cities (and on which we have focused in Part Two of this book) is the importance of mixing economic diversity into the urban fabric, through ensuring that there is good "morphological variety", that is, a range of affordable commercial spaces, in old and new buildings, of varying sizes and types. There is a need to avoid homogeneous residential areas in the centre of cities, and re-create what Jane Jacobs described as that 'most intricate and close-grained diversity of uses that give each other constant mutual support both economically and socially' [11, p. 19]. Examples of how older industrial buildings are being used to provide new commercial spaces in cities include the Spectacle Works in the old Jewellery Quarter in Birmingham, which has been converted into a live-work space for creative and craft industries, incorporating both apartments and studios. In Sheffield,

148 *Rebuilding Urban Complexity*

too, an old cutlery works called Portland Works has similarly provided a space for new business growth, hosting thirty-five small businesses including knife-makers, jewellers, guitar makers, bicycle makers, cabinet makers and a gin distillery [29],[7] while 150 artist and maker studios have been hosted at Persistent Works, Exchange Place and Manor Oaks [29, p. 79]. More recently, Leah's Yard, a historic building (which had consecutively hosted a manufacturer of shears, a silver die stamp maker and then a collection of small industrial workshops) has also been transformed into a centre for creative firms and independent retailers in Sheffield as part of the city's "Heart of the City II" project.

Invest in a new industrial urbanism: Policy makers in cities such as London and New York are now working to attract *manufacturers* back into the heart of their cities, taking advantage of the "incubating role" that cities can have for small companies – as opposed to relocating these manufacturers into peripheral business parks and estates. This is becoming much more acceptable now that manufacturing firms are generally often smaller and less polluting than they were in the past, and because they offer important jobs in middle-level skilled occupations which are often missing in increasingly polarised knowledge-based urban labour markets [30–32]. Many cities now see a deep divide between well-paid knowledge workers and much more poorly paid lower-skilled service workers that support them, with limited middle-skilled jobs in between. In New York, planners are seeking to protect the rich mixture of industrial and other uses in areas such as Long Island City and Gowanus. Hackney in London has also seen an innovative incorporation of manufacturing spaces in recent years into an urban fabric characterised more broadly by rising residential property prices and gentrification (see Box 11). This is supported by a broader strategy in London to have "no net loss" of manufacturing land following new property development [see 33].

In his book *Working Cities* [30, 34], the American architectural theorist Howard Davis cites these examples as representing a new "industrial urbanism", where manufacturing is woven back into everyday life in ways that support both local vitality and also meaningful work. Davis particularly emphasizes the importance of three factors in this type of urbanism – hybridity, "terroir" and visibility. Hybridity refers in part to combined living and working spaces – a use of space which is also championed by Francis Hollis [35], and which has become increasingly valued as "working from home" became a new norm for some parts of the labour force during and after the Covid-19 pandemic.[8] Davis is also referring, however, to 'hybrid neighbourhoods', where production is woven into the urban fabric in linear formations – industrial streets and ribbons as opposed to being segregated in industrial parks. Davis and I have written together about the benefits of lines of *railway arches* in this regard, as the modular and messy nature of these "residual" urban spaces offers opportunities for innovation and experimentation, such as we have seen in Manchester [36]. The UK's public sector lost an opportunity to actively manage such spaces for the good of inclusive entrepreneurship when they were privatised in 2018 [37]. More recently, the perils of insensitively developing railway arches as part of place-based regeneration were also made evident in Hackney (see the

Rebuilding urban complexity: how can policy makers intervene? 149

second part of Box 11). Nevertheless, local authorities could do well to take on the role of being more sensitive, hands-off stewards to these useful manufacturing and (recycling) spaces in the future, subsidising access to promote more inclusive entrepreneurship.

Box 11   Learning from Hackney in London

The "maker mile" in Hackney: a hybrid neighbourhood?

The borough of Hackney in London offers an example of how smaller-scale manufacturing and artisanal forms can still be hosted in a gentrifying area with relatively high property prices [see 33]. In London Fields, for example, buildings such as Regents Studios and Netil House have provided incubator spaces for a range of artisanal companies from embroiderers to ice cream makers. This area of London also hosts several clusters of shipping containers stacked on top of each other (such as Containerville and the Gossamy City Project), offering high-density bases for start-up companies, both artisans and customer-facing services. The local materials-based "making" economy appears to work symbiotically with the wider knowledge-based creativity which is strong in both Hackney and neighbouring Shoreditch. Local firms also frequently supplement their revenue by offering training courses to local people in how to work better with their hands – from making bread to wooden spoons and furniture.

The Morning Lane arches: a "case study of what not to do"?

Railway arches remain an important host for manufacturing and repair in Hackney [see 36]. However, a less positive example of urban regeneration in this borough is the renovation of the Morning Lane railway arches near Hackney Central, using regeneration funds which were meant to help the area address the conditions that had led to rioting in 2011. A group of 12 railway arches received a make-over to turn them into a high-end fashion outlet called Hackney Walk at a cost of £100 million. After a brief flourishing, the scheme failed, and these arches are now all boarded up and covered in graffiti [38]. This scheme was to some extent exploiting local "latent potentials", building on the fact that the streets around had already attracted fashion outlets such as Burberry and Aquascutum, while Mare Street also hosts a fashion college. However, the scheme backfired by deciding what these arches should be used for (relatively high-end outlet retail shops set behind new curated spaces fronted with gold-coloured metal and glass), as opposed to letting the uses emerge on the basis of providing to the market messier, less "curated" types of space.

150   *Rebuilding Urban Complexity*

By talking about "terroir" in *Working Cities*, Davis is referring to the added value of producing things that have an urban form of rootedness – an example being the many urban breweries that have flourished in the UK in recent years (not least under railway arches) through exploiting quirky local credentials. The idea of "visibility" is also an important concept emerging from Davis's work. By being able to see production, we become more likely to want to participate in it [see also 39], and also better able to ethically control it – in terms of both the human conditions of production and also its environmental consequences. Indeed, a key role for a new "industrial urbanism" should be the recycling and remanufacture of goods in lattice-style economies, as opposed to new lines of "virgin" production.

Consider spatial configuration when designing new retail developments and protect spaces which host an "everyday" economy: Small-scale retail will also continue to be important in breathing new life and economic diversity into post-industrial city streets. There are today many empty shops in our high streets as a result of fierce competition from not only e-commerce but also large shopping malls, both within and outside of city centres. As city planners in cities like Sheffield work to re-establish their central commercial cores, they should take into account the configuration of their city streets – in particular to identify where their "topological" global city centre might be (where there will be most footfall) as opposed to where areas may be more *locally* integrated and hence appropriate for other types of land use. Again, there is a complicated multiscale connectivity which underlies successful shopping cores in cities.

Many city centres in the UK are also exploring ways of preserving and reintroducing more diverse forms of "everyday" and independent retail in the face of competition from global brands – with Newcastle City Centre offering an example of how a "protected space", the Grainger covered market, is a valuable asset in such a fight (see Box 12).[9] Alleyways in the heart of cities also often provide valuable spaces for everyday goods to be bought and sold, and hence deserve protecting (with Reading, for example, hosting "Smelly Alley" – Union Street – in its centre, named due to the fishmongers which it hosted until recent times, in addition to everyday services such as greengrocers, bakers and locksmiths).

Box 12   "A protected space" – the Grainger covered market

The Grainger market was built in 1835 as part of the Georgian interventions in Newcastle city centre described in Chapter 7. It is a complex indoor niche which has been supporting small businesses and traders for nearly 200 years. The market is now owned and operated by Newcastle City Council and has over one hundred commercial units. While it started off hosting butchers and vegetable stalls, it now hosts a much more diverse range of sellers [40]. These include many retailers that form part of the "everyday economy",[10] providing goods and services that are more difficult to find elsewhere in the city centre – not only butchers and greengrocers but also plant stalls, bakers, small-scale independent traders of jewellery, clothing and books, and also cafés and bars. A new store offering "waste-free" shopping has opened up. A much older concern in the Grainger market is the "weigh house". Products were once weighed here; now *people* are weighed to keep track of their health. The council has a small room in the market from which it manages the space.

Overall, this market represents a valuable "protected space" – protected from the elements but also from the market forces which might otherwise have outcompeted all the valuable everyday retailers who are based there. It is also a protected piece of *heritage* and continuity in the heart of the city. The Grainger market was allocated Levelling Up funds to improve signage and conditions in the market, and to allow it to potentially stage live events in the future. Elsewhere, a covered market in Oxford has played a similar role in conserving many small businesses in the heart of a busy and highly competitive commercial centre.

Harness latent economic potentials: So far, we have been focusing on synergies between *spatial* and economic forms of complexity. Intervening spatially in cities will clearly have broader economic impacts – rebuilding urban spatial complexity in planning-disrupted cities will have knock-on effects on the broader prosperity of urban economies. Expanding the street-based fabric of urban cores will create more "generative" space for encounter and creativity – supporting the knowledge-based economy but also other urban economic sectors that benefit from agglomeration externalities and mutual accessibility. Reconnecting local areas into multiscale networks would better embed local economic clusters such as MediaCity into citywide knowledge-sharing networks. Creating more economically diverse neighbourhoods would create not only urban vitality but also potential new synergies between the knowledge economy and new forms of "industrial urbanism", not least when it comes to repurposing and recycling.

Nevertheless, it is also interesting to explore how policy makers could nurture complexity in their economies in other (non-spatial) ways – harnessing the latent potentials which complex urban economies offer (as we saw in Part Two of the book), while also "tilting the playing field" towards public goals such as greater social inclusion and environmental sustainability.

In his popular book *Triumph of the City*, Ed Glaeser argues that formerly industrial cities can only reinvent themselves by 'shedding the old industrial model completely' [41, p. 43]. Arguably Manchester has similarly 'done its best to shed its manufacturing past while aggressively seeking to manufacture a new urban future' [42, p. xiv] – with this reinvention mainly being focused on services, culture and the digital transformation. Research on green transitions, reviewed in Chapter 10, suggests however, that the *materials-based* capabilities and economic interdependences

152   *Rebuilding Urban Complexity*

which have been so important to Manchester's past are likely in the future to be important as a source of regeneration and new growth. As Hidalgo [43] says, such latent potentials in local economies 'reveal paths that may be easier to climb'.

Manchester's deeply embedded cross-sector expertise in materials production means that it can make important contributions to the infrastructure developments which are important to urban sustainability in the context of climate change. Graphene is already being used for new forms of "green tech", from water desalination to hydrogen storage. The city's capabilities in textiles, clothes, chemicals and gases also make it a strong candidate to become a global leader in sustainable textiles and clothing. This would be a welcome contrast to the current situation, where the city's textiles and clothing sector is often very much part of the environmental problem – producing, importing and exporting throwaway materials, cheap clothes and "fast fashion". The World Economic Forum has published a sobering statistic: while global clothing production has roughly doubled since 2000, 85 per cent of clothes go to landfill [44].

The idea of cities building on their latent economic potentials is not new – it already features in European Cohesion policy, with its strand of research and funding associated with "smart specialisation" [45]. Nevertheless, such approaches are stymied by a lack of local economic data on economic interdependences [46]. In order for local policy makers to fully take into account the evolutionary potential of their own very specific local urban complexity, we must become much better at producing and updating local datasets – something which may be helped in the future by artificial intelligence and machine learning. In a broader sense, it has been found that boosting digital skills across the economy is an important way of supporting broader green economic diversification [47]. As artificial intelligence, in particular, becomes more broadly important to our local labour markets, we will need to be aware of how to enhance the aspects of this technology that are *complementary* to both human skills and embedded local capabilities, rather than directly *substituting* the local jobs which sustain urban livelihoods.

Harness urban problem-solving to achieve sustainability: Indeed, new urban problem-solving does not need to occur within an overall framework of economic *growth and profitability*. While Mariana Mazzucato emphasises the importance of retrieving "public purpose", Danny Dorling's recent book *Slowdown* [48] suggests that we should be focusing on *stability* rather than growth in the context of a global deceleration associated with population decline and climate change. In this context the stable, ongoing complex niches which support everyday economic needs – Castle Street in Edgeley in Stockport and Grainger Market in Newcastle – are particularly valuable. Such thinking also suggests that the bottom-up self-organising and problem-solving characteristics of cities might be most usefully turned towards the many challenges that remain if we are to make urban living less unequal and more sustainable in the face of climate change. As Jane Jacobs explained, cities often in fact advance economically by solving their own problems – with the most successful cities having 'energy enough to carry over for problems and needs outside themselves' [11, p. 585].

Rebuilding urban complexity: how can policy makers intervene? 153

Restore pride through new forms of leadership: The idea that post-industrial cities can again *lead* social and economic change will be an important way of rebuilding pride – with these cities no longer being seen as "places left behind" or "places that don't matter" [see, e.g., 49] but rather the sites of new branches of human creativity. As Douglas Rae points out, 'people and institutions make commitments to places not at random, but on the basis of promise' [50, p. 42]. Cities in the UK have been devolved a series of additional powers in recent decades, as part of a series of UK "city deals", and an increased retention of business rates to spend locally. Formerly industrial cities such as Manchester have been hubs, not only of economic invention, but also of radical social change. This is the time for post-industrial cities to reradicalise – not to reimpose new forms of top-down spatial order, but rather to fight for a return to the urban complexity and integrity which once made them so successful – that 'marvellous order' of which Jane Jacobs spoke. At the same time, they will need to build a new relationship between this "marvellous urban order" and the fragile but equally self-organising ecological systems on which they depend.

Notes

1 The "survey before plan" approach had been followed by Ildefons Cerda in Barcelona, who spent many years observing how the city worked before developing his plan to extend the urban fabric beyond the former city walls in 1860 – which resulted in this new urban fabric being knitted very effectively into the street system of the historic core.
2 Despite their similarities of approach, Jacobs did not fully appreciate how Geddes parted in his ideas from the "garden cities" movement with which he became associated. See Batty, M., and S. Marshall, Thinking organic, acting civic: the paradox of planning for cities in evolution. *Landscape and Urban Planning*, 2017. **166**: p. 4–14.
3 I gained valuable insights into Mariana Mazzucato's approach when acting as an Honorary Research Fellow in her Institute for Innovation and Public Purpose at University College London.
4 With thanks to Ed Parham of Space Syntax for making these diagrams available. While this bid was not successful, the actual redevelopment of the Kings Cross site by Allies and Morrison and Demetri Porphyrios has provided a good example of how a relatively hands-off approach can lead to an integrated mixed-use niche in the city. Although there was no design code, "parameter plans" were set out which built on existing infrastructure (not only local streets but also the canal and historical leftovers such as the gasworks) and indicated to each new building development its 'duty to contribute to the greater whole' (see Moore, R., 'Nervous of its own boldness': the (almost) radical rebirth of King's Cross. *The Guardian*, 28 April 2024).
5 Such a facade on one side of the Westgate shopping centre has caused a problem in Oxford, for example, by restricting the use of this street, with a knock-on impact on the city's adjacent "Castle Quarter" which has seen heritage-based regeneration investment in recent years.
6 With thanks to Nalin Seneviratne for a fascinating discussion about recent urban design thinking in Sheffield.
7 www.portlandworks.co.uk/. Accessed [07/06/2024].
8 At the time of writing, a weekly calendar involving "part of the week in the office, part of the week at home" has become the new norm for workers who are not involved in front-line or face-to-face services – helping to also reshape the temporality of

154 *Rebuilding Urban Complexity*

commuting patterns in cities and the distance which people are willing to travel to work as commutes become less frequent.

9 I am grateful to my colleague Patricia Canelas for our recent discussions about "protected spaces" in cities and their stewardship.

10 The "everyday economy" is a topic of increasing interest amongst planners, being the topic for a research cluster which I participate in at the Bartlett School of Planning led by Jessica Ferm and colleagues. This is part of a broader interest today in the "foundational economy" – economic development which puts local infrastructure and quality of life ahead of profit-driven economic growth.

References

1. Meadows, D.H., *Thinking in systems: a primer*. 2008, White River Junction, VT: Chelsea Green Publishing.
2. Jacobs, J., *The economy of cities*. 1970, New York: Vintage Books.
3. OECD Global Science Forum, *Applications of complexity science for public policy: new tools for finding unanticipated consequences and unrealised opportunities*. 2009, Paris: OECD Publishing.
4. Batty, M., The size, scale and shape of cities. *Science*, 2008. **319**: p. 768–771.
5. Chandler, D., Beyond neoliberalism: resilience, the new art of governing complexity. *Resilience*, 2014. **2**: p. 47–63.
6. Lindblom, C., The science of "muddling through". In *Classic readings in urban planning*. 2018, London: Routledge. p. 31–40.
7. Rittel, H.W. and M.M. Webber, Dilemmas in a general theory of planning. *Policy Sciences*, 1973. **4**(2): p. 155–169.
8. Batty, M. and S. Marshall, Centenary paper: the evolution of cities: Geddes, Abercrombie and the new physicalism. *The Town Planning Review*, 2009: p. 551–574.
9. Bettencourt, L.M., Designing for complexity: the challenge to spatial design from sustainable human development in cities. *Technology| Architecture+ Design*, 2019. **3**(1): p. 24–32.
10. Batty, M. and S. Marshall, Thinking organic, acting civic: the paradox of planning for cities in evolution. *Landscape and Urban Planning*, 2017. **166**: p. 4–14.
11. Jacobs, J., *The death and life of great American cities [1993 edition]*. 1961, New York: The Modern Library.
12. Dovey, K., et al., *Atlas of informal settlement: understanding self-organized urban design*. 2023, London: Bloomsbury Publishing.
13. Cities Alliance., Cities without slums. In *Action plan for moving slum upgrading to scale*. 1999, Washington: Cities Alliance.
14. Echanove, M. and R. Srivastava, 'Mess is more' – value and shortcomings of the city's ad hoc development process. In *Routledge handbook of Asian cities*, 2023, London: Routledge. p. 280.
15. Neuman, M., Planning, governing, and the image of the city. *Journal of Planning Education and Research*, 1998. **18**(1): p. 61–71.
16. Wohl, S., From form to process: re-conceptualizing Lynch in light of complexity theory. *Urban Design International*, 2017. **22**: p. 303–317.
17. Talen, E., Design that enables diversity: the complications of a planning ideal. *Journal of Planning Literature*, 2006. **20**(3): p. 233–249.
18. Mazzucato, M., R. Kattel, and J. Ryan-Collins, Challenge-driven innovation policy: towards a new policy toolkit. *Journal of Industry, Competition and Trade*, 2020. **20**(2): p. 421–437.
19. Dixon, T.J. and M. Tewdwr-Jones, Urban futures: planning for city foresight and city visions. In *Urban futures*. 2021, Bristol: Policy Press. p. 1–16.
20. Froy, F. and S. Giguere, *Breaking out of policy silos*. 2010, Paris: OECD Publishing.

Rebuilding urban complexity: how can policy makers intervene? 155

21. Giguère, S. and F. Froy, *Flexible policy for more and better jobs.* 2009, Paris: OECD Publishing.
22. Farrell, T., *The city as a tangled bank: urban design versus urban evolution.* 2014, New York: John Wiley & Sons.
23. Steigerwald, B., City views. Urban studies legend Jane Jacobs on gentrification, the New Urbanism, and her legacy. *Reason Magazine.* 2001.
24. Southworth, M. and E. Ben-Joseph, *Streets and the shaping of towns and cities.* 2013, Washington, DC: Island Press.
25. Sennett, R., The public realm. In *Being urban: community, conflict and belonging in the Middle East,* S. Goldhill, Editor. 2020, London: Routledge. p. 35–58.
26. Meyerson, M. and E. Banfield, *Politics, planning, and the public interest: the case of public housing in Chicago.* 1955, New York: Free Press.
27. Wainwright, O., 'Like something out of the flintstones': the luxury flats causing fury in Folkestone. *The Guardian.* 2023, London.
28. Dixon, R., Folkestone's seafront has been transformed by art in the past decade. *The Guardian.* 2021, London.
29. Jones, M.J.J., *Sheffield at work: people and industries through the years.* 2018, Stroud: Amberley Publishing.
30. Davis, H., *Working cities: architecture, place and production.* 2020, Abingdon, Oxon and New York: Routledge.
31. Dellot, B., *The second age of small: understanding the economic impact of micro businesses.* 2015, London: Regional Studies Association.
32. Autor, D.H., Work of the past, work of the future. In *AEA papers and proceedings.* 2019. Nashville, TN: American Economic Association.
33. Domenech, T., F. Froy, and N. Palominos Ortega, *The maker-mile in East London: case study report.* 2019, Brussels: Cities of Making.
34. Froy, F., Review of *Working cities: architecture, place and production* (1st ed.) by Howard Davis, 2020. Abingdon, Oxon and New York: Routledge in *Journal of Urban Design,* 2023. **28**(2): p. 254–256.
35. Holliss, F., *Beyond live/work: the architecture of home-based work.* 2015, Abingdon, Oxon: Routledge.
36. Froy, F. and H. Davis, Pragmatic urbanism: London's railway arches and small-scale enterprise. *European Planning Studies,* 2017. **25**(11): p. 2076–2096.
37. Froy, F., The real issue behind Britain's railway arches being sold off: small businesses will suffer, and that would be a serious loss. *The Independent.* 2018.
38. Usborne, S., 'It was a case study for what not to do': the regeneration project that became a £100m luxury ghost town. *The Guardian.* 2023, London.
39. Froy, F., H. Davis, and A. Dhanani, *Can the organisation of commercial space in cities encourage creativity and 'self-generating' economic growth? A return to Jane Jacob's ideas.* 11th International Space Syntax Symposium. 2017, Lisbon.
40. Young, Y., *The Grainger Market: the people's history.* 2015, Newcastle: Tyne Bridge Publishing.
41. Glaeser, E., *Triumph of the city: how urban spaces make us human.* 2011, London: Pan Macmillan.
42. Peck, J. and K. Ward, *City of revolution: restructuring Manchester.* 2002, Manchester: Manchester University Press.
43. Hidalgo, C.A., *The policy implications of economic complexity. Research Policy,* 2023. 52(9): 104863.
44. McFall-Johnson, M., *These facts show how unsustainable the fashion industry is.* 31 January 2020, Geneva: World Economic Forum.
45. Foray, D., *Smart specialisation: opportunities and challenges for regional innovation policy.* 2015, Abingdon, Oxon and New York: Routledge.
46. McCann, P. and R. Ortega-Argilés, Smart specialisation, regional growth and applications to EU cohesion policy. *Working Document of the IEB,* 2011. **14**: p. 1–32.

156 Rebuilding Urban Complexity

47. Froy, F., et al., What drives the creation of green jobs, products and technologies in cities and regions? Insights from recent research on green industrial transitions. *Local Economy*, 2022. **37**(7): p. 584–601.
48. Dorling, D., *Slowdown: the end of the great acceleration—and why it's good for the planet, the economy, and our lives*. 2020, New Haven: Yale University Press.
49. Rodríguez-Pose, A., The revenge of the places that don't matter (and what to do about it). *Cambridge Journal of Regions, Economy and Society*, 2018. **11**(1): p. 189–209.
50. Rae, D.W., *City: urbanism and its end*. 2005, New Haven: Yale University Press.

Index

Note: Page numbers in *italic* indicate a figure or map on the corresponding page.

15-minute city 133, 144

accessibility, economic 50–51
active frontages 108, 116, 145–146; and
  blank facades 146
actor-network theory 16
affordances 7, 55–56, 66, 86–87, 93,
  99, 130
agglomeration theory 19, 48–49, 140;
  agglomeration externalities 20, 51–56,
  53, *56*, 59
Altrincham 42, 52, 58–59, *58*
artisanship 117, 119–121, 131

back streets 49, 87
biodiversity 128, 130, 133–136, 146
Bolton 22, 42, 51–52, 68, 70, 118, 129
boundaries 32–33, 40, *42*, *44*, 101,
  146
bounded rationality 142, 143
branching 6, 15, *67*, 72–73; chemicals
  68–69; and the dynamic division of
  labour 65–66; and differentiation
  within systems 15; economic 6, 65–76,
  67, 90–95, *91–92*; emergence and
  self-organisation 15, 80–82; engineering
  69–70; evolving open system 70–71;
  evolving spatial systems support
  branching 90–95, *91–92*; mapping
  morphological trajectories 82–90,
  83, *85*, *87–88*; resilience and decline
  73–76; textiles 67–68
buildings 16–17, 92–95, 98, 101, 103,
  106–113, 119–121, 146–149
Bury 22, 42, 55, 129
by-products 128–131

capabilities 5–6; and branching
  94–95; how firms use capabilities
  interchangeably 30–31; and green
  industrial transitions 131–132; loss of
  75; and parts and wholes 21, 29, 33; and
  policy makers 139–140, 151–152; and
  waterproofing 101, 103
centrality 39–40
chaos 10, 14, 49, 105–106, 108, 142;
  labyrinth 106
chemicals industry 24, 26–30, 66–70, 130,
  132, 140, 152
circular economy 128–131; *see also* lattice
  economies
cities 7–8; comparing local embeddedness
  in 31–32; dual system of city streets
  39; open cities and defining systems
  boundaries 32–33; as partially ordered
  and open spatial systems 40–41; spatial
  arrangement of industry relatedness in
  56–59, *57–58*; and structure 47–49,
  48; as systems of systems 127–136;
  Victorian 105–106
cities as open systems 14, 32–33, 40–41,
  105, 131, 140–141
climate change 16, 76, 128, 152
clothing 26–33, *28*, 67–69, 98–104,
  129–130, 140, 152
colonialism 13, 71, 102, 127
commercial building space 119–121
common problem definition 30
communities of production 25, 27–28
complexity: and contamination 130;
  emergence and self-organisation 80–82;
  evolving spatial systems support
  branching 90–95, *91–92*; grids 84–86;

158   *Index*

loss of 74–76; mapping morphological trajectories 82–90, *83, 85, 87–88*; and new policy directions 143–153, *145*; spatial 37–41; theories of 10–17; *see also* urban complexity

configurational analysis 47, 59–60; of local economies 20–33; mapping differential economic accessibility 50–51; space syntax of agglomeration externalities 51–56, *53, 56*; spatial arrangement of industry relatedness 56–59, *57–58*; and structure 47–49

connectivity 118

contamination 130

core, urban 41–42, 144; building out from 119; commercial building space 119–121

clusters 19–20, 48–50, 55–56, *56*, 59, 92–94, 102, 129, 151

culture, spatial 40, 82, 102–103

cycling 132–133

decline 7–8, 73–76, 103, 105–106, 144, 152

density 41, 47–49

design codes 82, 146–147

destruction of complexity 1; building "up" instead of "out" 119; impact of planning changes on industry 113–116; through imposition of a new order 106–118, *107, 109, 111, 114*; loss of commercial building space in city core 119–121; loss of network integrity 116–118; Victorian cities as sites of chaos and disorder 105–106; voids 118–119

Detroit 1, 90, 117, 127

development 6–7

differential economic accessibility 50–51

disorder 105–106

diverse local economies 20–21, 55–59, 86–91, 147–148

division of labour 65–66

dual system of streets 39, 43–45

East Kolkata Wetlands 135–136

economics: economic accessibility 50–51; economic communities 28–30; latent economic potentials 151–152; part-whole understandings 19–20

economic branching 6, 65–76, *67*, 90–95, *91–92*; chemicals 68–69; and the dynamic division of labour 65–66;

engineering 69–70; evolving open system 70–71; evolving spatial systems support branching 90–95, *91–92*; resilience and decline 73–76; textiles 67–68

economic geography 1, 6, 15, 20, 48; evolutionary economic geography 20, 66

economies 5–8; chemicals 68–69; circularity and recycling 128–131, 136, 150; comparing economic communities 28–30; complementary urban economies and the Northern Powerhouse 59; economic diversity and the dynamic division of labour 20–21, 65–66, 86–89, 90–92, 147–148; engineering 69–70; "everyday" 150; evolving open system 70–71; evolving spatial systems support branching 90–95, *91–92*; how firms use capabilities interchangeably 30–31; labour sharing 24; lattice 128–131; looking at the economy at a finer grain 24–28, *28*; mapping differential economic accessibility 50–51; open systems and defining systems boundaries 32–33; part-whole understandings 19–20; resilience and decline 73–76; space syntax of agglomeration externalities 51–56, *53, 56, 57*; spatial arrangement of industry relatedness 56–59, *58*; and structure 47–49, *48*; supply chain relationships 22–24, *23*; textiles 67–68; topological structure 20–21

embeddedness, local 29–32

emergence 6, 13, 15, 73, 80–82, 105, 140, 143–153

Engels, Friedrich 105–106

engineering 26–29, 68–72, 140

"everyday" economy 150

evolving 15–16, *67*, 72–73; chemicals 68–69; and the dynamic division of labour 65–66; emergence and self-organisation 80–95; engineering 69–70; evolving open system 70–71; evolving spatial systems support economic branching 90–95, *91–92*; mapping morphological trajectories 82–90, *83, 85, 87–88*; resilience and decline 73–76; and spatial complexity 90–95, *91–92*; textiles 67–68

expansion 89–90;

Index 159

exploitation 71, 76, 102, 127; and colonialism 71, 102, 127; and slavery 71
externalities, agglomeration 20, 51–56, *53, 56, 57*

facades 146
Folkestone 147
foreground network 39, 43–45, *44*, 49, *57*, 81–89, 102, 106, 117, 139; parts and wholes 39, 43–45, *44*; and policymakers 144–145; and spatial location of different economic sectors 50–51
"fossil capital" 76, 128, 136

geographies of extraction 71
Gephi 23
global production chains 32–33
global supply chains 55–56; *see also* supply chains
GOAD maps 51, 86–88, *87, 88,* 129
Grainger market 150–151
graph theory 7–8, 21, 37–38, 45n4
Greater Manchester *see* Manchester
green industrial transitions 131–132
grids 81, 84–92, *85, 87, 88,* 145–146

Hackney 148–149
Hanson, Julienne 37–39, 75, 82, 108, 117
hierarchy 13–14, 25–26, *26*
Hillier, Bill 6–7; and common problem definition 30; and constraints relating to liveable urban systems 40–41; and eco-villages 133; and the evolution of street systems 80–81, 86; and form and function 91; and partially ordered systems 49; and perpetual night syndrome 108; and policy makers 144–145; and space syntax 37–41
hybrid neighbourhood 148–149

industrial transitions 131–132
industry relatedness 21, 56–59, 66, 98
industry/industries: green industrial transitions 131–132; impact of planning changes on 113–116; industrial urbanism 148–150; intermingling of 86–89; left-over industrial land and contamination 118–119, 130
inefficiency 16, 94–95, 110, 115
informal settlements 81, 142

infrastructure 16–17, 143–144; grids as infrastructure 84–86; nature-based 135–136; transport infrastructure 90
inner ring 41, 112
innovation 52–53, 65–66, 70–74, 94, 99–101, 129; green innovation 131–132
integration, spatial 41–43
integrity 17, 107, 112, 116–118, 130, 142
interventions 139–140; circling back to policy 140–143; new policy directions 143–153, *145*

Jacobs, Jane 6–7, 53, 55, 80, 88, 94; and branching 65–66, 73, 75, 99, 104; and the destruction of urban complexity 75, 110; and "eyes on the street" 134, 145, 147; and parts and wholes 20, 36, 39–40; and policy makers 141–142, 144–145, 147, 152; and theories of complexity 12, 15; and urban recycling 129;
Jay Trim 56, 60, 115–116

Kauffman, Stuart 10, 15, 66, 99
knowledge sharing 52–54, 58–59, 151

labour: division of 65–66; labour sharing 24, 52, *53*
labour market 47, 52, 102, 148, 152
latent potential 6, 8, 15–16, 140, 149, 151–152
lattice economy 1, 128–131, 136, 150
leadership 153
left-over industrial land 118–119, 146
Levelling Up 7, 112, 151
local development 6–7
local economies *67*, 72–73; chemicals 68–69; and the dynamic division of labour 65–66; engineering 69–70; evolving open system 70–71; resilience and decline 73–76; textiles 67–68; *see also* economies
local embeddedness 29–32
local policies 5–6
local scale: intermingling of industries 86–89
loss of complexity 74–76

macintosh 98–101, *100*
Macintosh, Charles 98, 101–102
Manchester 7–8; a 'city remaking itself out of its own history' 66–73, *67*; applying relatedness models to 22–33, *23, 25, 27, 28*; and the destruction

160   *Index*

of urban complexity 106–111, 114–121; and the lattice economy 128–131; the Manchester hive 47; the morphological trajectory of 82–90; and nature 134–135; policy makers 139–140, 143–148, 151–153; spatial configuration of 41–44, *42–44*; spatial organisation of economic activities 50–59; and waterproofing 98–104

manufacturing 30, 49, 51, 115–121, 151: and decline 74–76; and the destruction of urban complexity 105–106, 113–121; and industrial urbanism 148–150; and messy spaces 94; and policy makers 148–149; and relatedness 24, 26–32

mapping 58–59, *58*; differential economic accessibility 50–51; morphological trajectories of post-industrial cities 82–90, *83, 85, 87–88*, the spatial configuration of cities 37–45, *42–45*

market-based rationalities 141–142

Marshall, Alfred 19–21, 70, 72

master plans 6, 38, 80, 82, 141, 146–147

materiality 16–17

MediaCity 118–119, 121

messy spaces 94–95, 120, 148

metalworking 31, 70, 72–73, 103, 127, 140

Morning Lane 149

morphological trajectories of post-industrial cities 82–90, *83, 85, 87–88*

national connectivity 118

nature 127–136

nature-based infrastructure 135–136

Neffke, Frank 7, 21, 24–25

networks: foreground 43–45, 145; loss of network integrity 116–118; and textiles 68; visualising 23

Newcastle 73, 82–86, 89–90, 95, 110–114, *113–114*, 117, 150–151

New Haven 51, 85, 89–92, 105–106, 112–113, 127

"new jobs from old" 104

niches 11, 40, 49, 55, 66, 93, 130, 147, 150

Northern Powerhouse 7, 59

Northern Quarter 42, 51, 84; historical economic intermingling 86–89

open system: cities as partially ordered and open spatial systems 40–41; defining systems boundaries 32–33; evolving 70–71

order 40–41, 105–121, *107, 109, 111, 114*; as compared with structure 38

outdoor leisure trade 103

Oxford Road Corridor 53, 119

partially ordered systems 14–15, 40–41, 49

parts and wholes 11–17, 19–20, 36–37, 144; comparing economic communities 28–30; looking at the economy at a finer grain 24–28, *25–26, 28*; open systems and defining systems boundaries 32–33; topological structure 20–21

path dependency (and path interdependency) 103, 128, 131, 136, 141; path interdependency 66, 71, 76, 95, 98, 103, 128, 140 143

planning 80–82, 105–116, 140–143; and industry 113–116; post-war 107–118; top-down 1, 80, 106, 121

policy 5–6, 139–153; circling back to policy 140–143; new directions 143–153, *145*

"pop-up" urbanism 146–147

post-industrial cities 7–8, 105, 116–118, 130, 135–136; morphological trajectories of 82–90, *83, 85, 87–88*; new policy directions 143–153, *145*

Private White V.C. 31, 33, 60, 74, 100–101

problem-solving 66, 73, 94, 98, 127, 130, 136n1, 152

production chains 30–33

railway arches 32–33, 94, 148–150

rationality, bounded 142–143

redundancy 16, 94–95, 110

regeneration 108–110, 115, 119, 147–149, 152–153

regional connectivity 118

relatedness models 22–33, *23, 25, 28*

relational 11–12, 20, 33, 55

relationships, topological 12

resilience 8, 15–16, 73–76, 115–117, 139; economic resilience 21, 30

retail developments 150

retraction 74–76

ring roads 44, 107–112, *107*, 118, 143

Salford 30–31, 82–84, *120*

self-organisation 1, 6–7, 13, 19, 80–82, 94, 110, 116, 119, 130, 141–142

Sheffield 19, 31: and branching economies 72–76, 80–86, 89–90; and the destruction of urban complexity 105–108, 110–112, 116–119; economic relatedness analysis 59, 72–73; and green industrial transitions 131; and outdoor leisure industry associated with the Pennines 103; parts and wholes 40; and policy makers 140, 143–150; spatial configuration of 43–45

size of cities 47–49, 133

skills basins 25–27, *27*, *28*, 31

skills relatedness 25–33, *25*, *27*, *28*, 33n1

slavery 71, 73, 128

space: adapting to urban spatial change 115–116; configuration of urban space 36–45; generic properties of 91–92; loss of commercial building space 119–121; mapping differential economic accessibility 50–51; space syntax of agglomeration externalities 51–56, *53*, *56*; spatial arrangement of industry relatedness in a city 56–59, *57*; spatial culture 82, 102–103; spatial integration 41–43, *42*, 51; spatial organisation of waterproofing 101–103; spatial "reach" 102; and structure 47–49; *see also* urban space

space syntax 37–41; of agglomeration externalities 51–56, *53*, *56*; and mapping differential economic accessibility 50–51; and the spatial evolution of cities 80–81

spatial complexity 37–41; emergence and self-organisation 80–82; evolving spatial systems support branching 90–95, *91–92*; mapping morphological trajectories 82–90, *83*, *85*, *87–88*

spatial configuration 36–45, 150; Greater Manchester 41–45, *42–45*; Sheffield 43–45

spatial culture 82, 102–103

spatial integration 41–43, *42*, 51

spatial segregation 41–43, 55, 81, 83, 116, 148

spatial systems: cities as partially ordered and open spatial systems 40–41; economic branching and evolution 90–95, *91–92*

Stockport 22, 42, 49–51, 86, 129, 152

Strangeways 49, 55–57, *56–57*, 100, 102–103, 115

streets/street systems 7, 17, 36–45, 145–146; dual system of city streets 39, 43–45; emergence and self-organisation in 80–82, compared with new urban forms such as "walkways in the sky" 107–113

structure 13, 47–49; as compared with order 38; topological structure of diverse local economies 20–21

supply chains 22–24, 27–28; urban "vantage points" into 55–57, *57*

sustainability 128, 151–152

systems theory 8, 10–17, *12*, 65; *see also* systems/systems thinking

systems of systems 127–136

systems/systems thinking 10–17; dual system of city streets 39; emergence and self-organisation in urban street systems 80–82; learn "systems lessons" from history 144; and nature 127–136; open systems defining systems boundaries 32–33; *see also* open systems; spatial systems; systems theory

textiles: and branching 66–75; and fashion in Greater Manchester 54–56, *56*; industry relatedness associated with textiles and clothing 26–30, *28*; and parts and wholes 19–24, 26, *28*; and policy makers 140, 152; and recycling/reuse of waste 129–130; waterproofing 98–104

theories of complexity 10–17

topology/topological structure 12–13, 26, 36, 44, 57, 65; of diverse local economies 20–21; topological centres of cities 84, 150

Tottenham Court Road 49, 92–93

trading interface 82–84

unplanned/informal settlements 81, 142

urban complexity 1, 5–8, 139–140; policy considerations 140–153; *see also* destruction of urban complexity

urban economies: comparing economic communities 28–30; how firms use capabilities interchangeably 30–31; labour sharing 24; looking at the economy at a finer grain 24–28, *28*; open systems and defining systems boundaries 32–33; part-whole understandings 19–20; supply chain relationships 22–24, *23*; topological structure 20–21; *see also* economies

urban fabric 22, 38, 42–44, 51, 83, 86, 89–90, 109, *109*, 112, 116, 118, 144–146; being knitted into the urban fabric 55, 142, 145, *145*; and walking/cycling 132–133

urbanism 146–147; industrial 148–150

urban problem-solving 152

urban space 36–37; space syntax 37–41; spatial configuration of Greater Manchester 41–45, *42–45*; *see also* spatial configuration

162 *Index*

vantage points 38, 55–57, *57*, 140
Victorian cities 105–106; creative use of by-products in 128–131
voids (industrial voids) 118, 121, 144–145

walkability/walking 132–133, 146
"walkways in the sky" 107–113
waterproofing/waterproofs 98–104, *100*

wholes *see* parts and wholes
wholesalers and merchants 55–56, 70–71, 115–116
Wright Bower 30–33, 60, 74

Xpose 54, 60

zoning 113–116